Neon Gods of the Afterworld

How We Can Reclaim Meaning, Power, and Humanity in a Digitally Engineered Age of Illusion

By Laura Lee

Lauxon Publishing, LLC

Published by Lauxon Publishing, LLC

For permission requests, please contact:
Email: info@Lauxonpublishing.com
Website: www.Lauxonpublishing.com

Author's Note

This book was written from a place of observation, reflection, and hope. It is not meant to demonize innovation or romanticize the past. Technology has brought extraordinary progress—connecting families, advancing medicine, expanding education, and empowering creativity in ways that were once unimaginable.

At the same time, I believe we are at a cultural turning point.

The systems that shape digital life are powerful. They influence attention, behavior, relationships, and even democratic processes. Many of these systems were optimized for growth, engagement, and efficiency. Fewer were designed explicitly for human flourishing.

This book asks a simple but profound question:
What would happen if we optimized for well-being instead?

The ideas within these pages are not rigid conclusions—they are invitations to dialogue. My intention is not to prescribe a single solution, but to encourage awareness, responsibility, and conscious participation in shaping our shared digital future.

We are not powerless inside these systems. We are participants in them. And participants can influence direction.

My hope is that this work sparks reflection—at the individual level, within families, inside organizations, and across communities. If it inspires even one person to reclaim attention, strengthen relationships, design ethically, or engage more thoughtfully with technology, then it has fulfilled its purpose.

Thank you for reading with openness and curiosity.

Table of Contents

Part II – The Psychological Afterworld

Chapter 9: Dopamine, Design, and Dependency
Platform architecture is engineered to hijack our neurological reward systems.

Chapter 10: Anxiety in the Age of Comparison
Constant exposure to curated lives intensifies insecurity and self-doubt.

Chapter 11: The Loneliness Paradox
We are more connected than ever yet increasingly isolated in meaningful ways.

Chapter 12: Digital Tribalism
Online communities amplify extremity and suppress nuance.

Chapter 13: The Performance of Self
We have confused visibility with worth and metrics with meaning.

Chapter 14: Burnout in a 24/7 World
Perpetual connectivity has erased the boundary between rest and responsibility.

Chapter 15: The Attention Crisis
Fragmented focus undermines creativity, learning, and deep thinking.

Chapter 16: The Infantilization of Society
Short-form stimulation is reshaping adult cognition and patience.

Part III – Power, Politics, and the Machine

Chapter 17: Surveillance as Business Model
Mass data extraction has normalized corporate oversight of private life.

Prologue

We live in a time of extraordinary connection and quiet fragmentation. Our devices glow with possibility, yet they also shape our attention, emotions, relationships, and beliefs in ways we rarely pause to examine. The systems we interact with daily were designed with remarkable intelligence—but not always with our well-being as their primary measure of success.

This book is not a rejection of technology. It is an invitation to see it clearly.

The neon glow of the digital world represents more than screens and software. It represents incentives. It represents design. It represents the unseen architecture that guides what we consume, how we think, and how we relate to one another. We did not wake up one day in this environment—it evolved gradually, shaped by innovation, competition, convenience, and human desire.

The purpose of this work is awareness.

When we understand how systems influence attention, identity, economics, education, and culture, we regain the power to choose. Awareness is not fear. It is freedom. It transforms automatic participation into intentional living.

This journey moves from recognition to restoration. It examines the structures that shape our digital lives—and then explores how we can rebuild balance, depth, community, and shared truth. The goal is not nostalgia. It is integration.

Technology will continue to evolve. The question is whether we will evolve with it consciously.

The future belongs not to the loudest systems, but to the most thoughtful participants.

This is the invitation: to see clearly, to think deeply, and to move forward deliberately.

CHAPTER . 1

The Birth of the Digital Deities

We did not set out to build gods. We set out to build tools. We wanted faster communication, easier access to knowledge, and more efficient ways to conduct business. The early architects of the internet imagined an open frontier of information where ideas could move freely and innovation could flourish. What they created instead was the foundation for systems that would quietly evolve beyond their original intent. These systems would not merely serve us; they would begin to shape us.

In the beginning, the digital world felt neutral. A search engine answered questions. A social network connected friends. An online store delivered convenience. Each innovation appeared as a solution to a specific human frustration. But as these platforms scaled, they began collecting data, studying behavior, and optimizing interaction. What started as utility slowly transformed into influence.

Influence, unlike utility, changes the direction of human thought. The moment platforms began learning from our clicks, pauses, and preferences, they began predicting us. Prediction led to personalization, and personalization led to subtle behavioral steering. Our feeds were no longer chronological reflections of reality. They became algorithmic interpretations of who we were and what would keep us engaged. We mistook this tailoring for service when it was, in fact, the beginning of governance.

Governance once belonged to institutions we could name and hold accountable. Governments passed laws. Religious leaders shaped morality. Communities reinforced norms. Now invisible systems guided our attention with mathematical precision. Their authority was not declared; it was embedded in code. And because code feels mechanical rather than moral, we failed to recognize its power.

The digital deities were not born in a single moment. They emerged gradually from the convergence of data, capital, and behavioral science. Venture funding accelerated growth at all costs. Behavioral psychology refined engagement tactics. Machine learning amplified feedback loops. Profit incentives aligned with attention capture. Before long, platforms were not just reflecting culture; they were actively constructing it.

We gave these systems access to our thoughts, relationships, and routines. We carried them in our pockets and placed them beside our beds. They became the first thing we consulted in the morning and the last thing we touched at night. Over time, they gained unprecedented intimacy with our habits. They knew when we felt lonely, anxious, curious, or bored. No deity in history has had such detailed knowledge of its followers.

Ancient gods demanded ritual. The new ones require interaction. A swipe, a like, a share, a scroll—these are the gestures of modern devotion. Each action feeds a system that grows more intelligent and more persuasive. The exchange appears harmless because it feels voluntary. Yet every interaction strengthens the architecture shaping the next decision. Worship no longer requires belief; it requires participation.

The most profound shift occurred when algorithms began mediating reality itself. News stories were ranked. Friendships were filtered. Advertisements were micro-targeted. Information ceased to be evenly distributed. Instead, it became curated, ranked, and optimized for engagement. Reality fractured into personalized streams.

Personalization feels empowering at first. It reduces noise and increases relevance. It gives the impression that the world understands us. But personalization also narrows exposure. It reinforces existing beliefs and shields us from discomfort. It turns the vast complexity of humanity into a mirror reflecting our own preferences. Over time, the mirror becomes a cage.

Unlike traditional gods, digital deities do not promise salvation in an afterlife. They promise validation in the present. Metrics replace meaning. Notifications replace affirmation. Followers replace community. The feedback loop becomes a system of micro-rewards that subtly rewires motivation. We begin seeking what performs rather than what is true.

The transformation was not malicious in its inception. Engineers optimized for growth. Designers optimized for engagement. Executives optimized for revenue. Each optimization seemed rational in isolation. But together, they created a system that prioritized attention above all else. Attention became the lifeblood of the new order.

Once attention became currency, competition intensified. Platforms refined their algorithms to hold users longer. Autoplay removed friction. Infinite scroll eliminated stopping cues. Notifications exploited psychological triggers. The goal was not satisfaction but continuation. A satisfied user logs off; an engaged user stays.

The birth of digital deities was sealed when artificial intelligence began adapting in real time. Systems no longer relied solely on static rules. They learned dynamically from billions of interactions. This learning enabled increasingly precise predictions. Precision enabled increasingly persuasive design. And persuasion, at scale, altered collective behavior.

We began outsourcing memory to cloud storage. We outsourced navigation to GPS. We outsourced social planning to event platforms. Gradually, we outsourced discernment to recommendation engines. What to watch, what to read, whom to follow, what to believe—suggestions filled the gaps once occupied by deliberate choice. Convenience slowly displaced contemplation.

The danger of subtle influence lies in its invisibility. Coercion is obvious and often resisted. Nudging is gentle and rarely noticed. Algorithmic nudges accumulate over time. They shape tastes, attitudes, and even political leanings. Because the process is incremental, the transformation feels organic. We rarely sense the external hand guiding the arc.

Power used to require physical force or institutional authority. Now it requires access to behavioral data and predictive modeling. The companies that mastered these capabilities became some of the most powerful entities in history. Their influence crossed borders without armies. Their platforms mediated billions of relationships daily. Their decisions shaped global discourse.

We call them companies, but their reach resembles mythology. They see across continents in milliseconds. They process more information in seconds than previous civilizations could in centuries. They influence elections, markets, and social movements. And yet their mechanisms remain largely opaque to those who rely on them. We interact with interfaces, not intentions.

The metaphor of gods is not hyperbole but illustration. Gods shape worldviews. They define what is rewarded and what is punished. They influence morality and aspiration. Digital systems now fulfill similar roles. They elevate certain voices and bury others. They amplify specific narratives while suppressing alternatives.

What makes these deities neon is not just their digital nature but their constant illumination. They glow from screens in darkened rooms. They pulse with notifications. They shimmer with curated imagery. They promise connection, knowledge, and belonging. Their radiance makes it difficult to look away.

We did not consent to worship in any formal sense. We simply accepted terms and conditions. We clicked agree without reading the covenant. In exchange for access, we surrendered data. In exchange for convenience, we surrendered privacy. In exchange for connection, we surrendered autonomy incrementally. The trade felt trivial at the time.

The birth of the digital deities reveals a broader human pattern. We create tools to solve problems. Those tools reshape behavior. Behavior reshapes culture. Culture reshapes identity. Eventually, we find ourselves living inside systems we barely understand. The cycle repeats throughout history, but never at this velocity.

Velocity is the defining feature of this era. Technological evolution once unfolded across generations. Now it unfolds across product cycles. Children adapt to platforms that did not exist when they were born. Norms shift before societies can evaluate consequences. Ethical reflection lags behind innovation.

Yet birth is not destiny. Recognizing the emergence of digital deities does not require rejecting technology. It requires understanding the incentives embedded within it. Systems behave according to their design. If engagement is rewarded, engagement will be maximized. If well-being is measured, well-being can be prioritized. Awareness is the first act of reclamation.

To confront the afterworld we now inhabit, we must first understand how it began. The birth of the digital deities was not a dramatic rebellion against humanity. It was a gradual transfer of influence from human judgment to algorithmic optimization. We handed over small decisions until the aggregate became profound. We mistook efficiency for neutrality.

This chapter is not an indictment of innovation but an invitation to examination. The systems we built are powerful because they reflect our own values amplified by code. If profit eclipses purpose, the outcome mirrors that imbalance. If speed outruns wisdom, instability follows. The architecture of the afterworld is not accidental.

The question before us is not whether technology should exist. It is whether we will remain conscious participants in its evolution. The digital deities thrive on passive devotion. They weaken under informed scrutiny. Their power depends on our unexamined habits. And habits can be changed.

The birth has already occurred. The systems are operational. The glow of neon fills our nights. But divinity requires belief to sustain authority. When we recognize influence, we regain choice. When we analyze design, we reclaim agency. The first step in reshaping the afterworld is understanding how its gods were made.

CHAPTER . 2

Worship in the Age of Algorithms

Worship has always required attention. Long before temples were built of stone, attention itself was the offering that signaled reverence. To focus on something is to elevate it above competing stimuli. In ancient societies, rituals organized collective focus toward shared symbols and stories. Today, our rituals are quieter but more constant. They unfold in the glow of handheld screens.

The algorithm is the unseen priest of the digital age. It decides what appears before us and what fades into obscurity. Unlike human intermediaries, it does not speak in moral language. It speaks in probabilities, correlations, and engagement metrics. Yet its decisions carry moral weight because they influence what we see as normal, desirable, or urgent. What it amplifies shapes what we value.

Every time we unlock a device, we participate in a ritual exchange. We offer attention, and in return we receive stimulation. The stimulation may come in the form of news, entertainment, affirmation, or outrage. The transaction feels voluntary and even pleasurable. But it is also structured. It is governed by systems designed to maximize duration and intensity.

Attention has become the primary commodity of the digital economy. In earlier industrial eras, raw materials fueled growth. In the information age, human focus is the scarce resource. Platforms compete not for money directly but for time. The longer we remain engaged, the more profitable we become. Our gaze is harvested, packaged, and sold.

The language of platforms reveals this shift. Users are described as "active daily users," "engagement rates," and "retention metrics." These terms reduce complex human beings to behavioral units. Each click is data. Each pause is measurable. Each scroll is monetizable. In this environment, reverence is quantified.

Worship in the age of algorithms does not require conscious belief. It requires habit. The habit of checking notifications. The habit of seeking updates. The habit of refreshing feeds for something new. These micro-rituals structure our days in ways we rarely question.

Consider how reflexive the gesture has become. A moment of boredom triggers a reach toward the phone. A lull in conversation prompts a glance downward. An uncomfortable emotion invites digital distraction. The device becomes both refuge and reflex. Over time, the reflex becomes dependency.

Algorithms are not inherently malicious. They are optimization engines. They analyze patterns to predict what will hold attention. If outrage keeps users engaged longer than nuance, outrage is amplified. If sensational headlines outperform balanced reporting, sensationalism spreads. The system learns from behavior without judging its impact.

This creates a feedback loop between human impulse and machine reinforcement. We are drawn to emotionally charged content. The algorithm detects this preference and supplies more of it. Our exposure increases our sensitivity and reaction. The cycle intensifies without central coordination. Worship becomes automated.

In religious traditions, rituals are often periodic. There are specific days of observance and defined times of prayer. Digital rituals are continuous. There is no sabbath built into the architecture of social media. The feed never ends. Notifications never sleep. The temple is always open.

The infinite nature of the feed alters our perception of completion. In physical spaces, books have final pages and conversations reach natural conclusions. Online spaces remove these boundaries. There is always another post, another video, another headline. The absence of stopping cues prolongs engagement beyond intention.

Wc tcll oursclvcs wc arc choosing what to consumc. Yct thc ordcr of presentation shapes perception. Items placed at the top of a feed receive disproportionate attention. Recommendations subtly guide curiosity. Trending topics imply collective importance. The algorithm does not dictate belief, but it frames relevance.

Framing is powerful because it operates beneath conscious awareness. When certain narratives dominate visibility, they begin to feel representative. When opposing perspectives rarely appear, they seem fringe. Over time, perception shifts. What is common becomes assumed. What is amplified becomes authoritative.

Worship in this context means deferring judgment to ranking systems. We allow likes and shares to signal credibility. We equate popularity with truth. We assume virality indicates significance. The algorithm becomes an arbiter of importance without declaring itself as such.

This deference reshapes identity. We begin tailoring our expressions to what performs well. We monitor feedback as a measure of worth. Posts that receive affirmation reinforce specific behaviors. Posts that are ignored discourage experimentation. The algorithm quietly conditions self-presentation.

There is a psychological comfort in being seen. Notifications provide small bursts of recognition. Each alert signals that we matter to someone or something. This intermittent reinforcement mirrors classic behavioral conditioning patterns. Unpredictable rewards are especially compelling.

The platforms understand this dynamic intimately. Designers test variations in color, sound, and timing to optimize response. A red notification badge triggers urgency. A subtle vibration signals relevance. These cues are not random. They are calibrated.

The more time we spend in algorithmically curated spaces, the more they shape our emotional rhythms. High-arousal content elevates stress levels. Rapid shifts between topics fragment attention. Constant comparison influences self-esteem. The architecture of the feed becomes the architecture of the mind.

Worship implies trust. Even if unconscious, participation reflects confidence that the system serves us. Yet the incentives driving algorithms are primarily commercial. Revenue often depends on advertising impressions. Advertising depends on sustained attention. Sustained attention depends on emotional engagement.

Emotional engagement is easiest to generate through extremes. Joy, fear, anger, and desire outperform moderation. Nuance requires patience. Patience reduces scrolling speed. Slower scrolling reduces exposure to advertisements. The economic model subtly favors intensity.

This does not mean every algorithmic interaction is harmful. Many people find community, education, and inspiration online. The issue lies not in isolated benefits but in systemic design. When profit aligns with psychological capture, the structure tilts toward excess. Balance becomes incidental rather than intentional.

The altar of technology is portable. It sits on desks, rests in pockets, and glows on nightstands. We carry it into meetings and meals. It interrupts conversations and infiltrates solitude. Its presence reshapes social norms.

Moments once reserved for reflection are now filled with input. Waiting in line becomes scrolling time. Quiet evenings become streaming sessions. The tolerance for silence decreases. The need for stimulation increases.

Children absorb these patterns early. They witness adults prioritizing screens over eye contact. They learn that attention is fragmented and externalized. Norms shift subtly across generations. The sacred space of undivided presence becomes rare.

To recognize worship is not to condemn technology entirely. It is to acknowledge imbalance. Attention is a finite resource. When it is perpetually diverted outward, inner life weakens. Reflection requires sustained focus.

If attention is the offering, then intention must become the safeguard. Without conscious boundaries, the algorithm defines the rhythm of our days. With awareness, we can begin reclaiming agency. We can decide when to engage and when to disconnect.

The age of algorithms is not ending. Their influence will likely deepen as artificial intelligence advances. The question is whether we remain passive participants or active stewards. Worship without awareness breeds dependency. Engagement with discernment cultivates autonomy.

To step back from reflexive scrolling is an act of quiet rebellion. To question why certain content appears is an act of cognitive sovereignty. To limit exposure is an act of self-preservation. These choices may seem small, but collectively they reshape habits.

Worship in the digital era thrives on unconscious repetition. Breaking the cycle requires deliberate interruption. It requires moments of pause before engagement. It requires remembering that algorithms serve corporate incentives, not personal flourishing.

The altar is luminous but not omnipotent. Its power depends on our participation. When we reclaim attention as a deliberate gift rather than an automatic offering, the balance begins to shift. The ritual can be redesigned.

The age of algorithms need not define our destiny. But unless we examine the nature of our devotion, we risk mistaking engineered compulsion for genuine desire. Awareness is the first crack in the illusion of inevitability. And from that crack, autonomy can reemerge.

CHAPTER . 3

The Currency of Influence

Gold once determined power. Nations fortified reserves to stabilize economies and project dominance. Wealth was tangible, heavy, and finite. Control over land and natural resources dictated global hierarchies. Today, the most valuable resource cannot be held in a vault.

Data has replaced gold as the dominant currency of influence. It flows invisibly through fiber-optic cables and wireless networks. It accumulates not in treasuries but in servers. It is extracted not from mines but from behavior. Every interaction becomes a deposit.

Unlike gold, data multiplies when used. Each search query refines predictive models. Each online purchase sharpens consumer profiles. Each location ping reveals movement patterns. The more data collected, the more accurate the predictions. Accuracy increases influence.

Influence is more powerful than ownership because it shapes choice. When a system predicts what you are likely to buy, watch, or believe, it can subtly steer outcomes. Steering does not require coercion. It requires positioning options strategically. Influence works best when it feels like preference.

The modern economy is built on behavioral forecasting. Corporations compete to anticipate desire before it becomes conscious. Recommendation engines suggest products before we articulate need. News feeds deliver stories aligned with prior engagement. Entertainment platforms curate content tailored to mood.

This predictive infrastructure rests on continuous surveillance. Surveillance in this context is normalized and often voluntary. We accept cookies. We grant permissions. We trade location access for convenience. The exchange appears harmless because the cost is intangible.

The true cost lies in asymmetry. Corporations know extraordinary detail about users. Users know little about corporate algorithms. This imbalance creates a power differential. Those who possess data can shape markets and narratives.

Markets now respond to behavioral analytics in real time. Dynamic pricing adjusts based on browsing history. Targeted advertisements follow users across platforms. Political campaigns micro-target messaging to specific demographics. The line between persuasion and manipulation blurs.

Data brokers operate largely out of public view. They aggregate information from multiple sources to build comprehensive profiles. These profiles include purchasing habits, social connections, and inferred interests. The individual becomes a composite of probabilities. Identity is reduced to predictive value.

The predictive value of a user determines advertising revenue. The more precisely a platform can forecast response, the more advertisers are willing to pay. This incentivizes deeper data collection. Depth increases predictive accuracy. Accuracy increases profitability.

Unlike traditional currency, data accumulates without diminishing the original holder's access. Companies can analyze user behavior repeatedly without depleting the resource. The scale becomes exponential. Billions of users generate trillions of data points daily.

The architecture of influence extends beyond commerce. Streaming platforms influence cultural trends by promoting specific genres. Social media platforms influence discourse by ranking posts. Search engines influence knowledge acquisition by prioritizing certain sources. Visibility becomes power.

Power once required physical infrastructure. Now it requires digital infrastructure. Server farms replace fortresses. Algorithms replace gatekeepers. Code replaces bureaucracy. The entities controlling these systems wield global reach.

Influence operates subtly because it aligns with convenience. We appreciate personalized recommendations. We enjoy curated playlists. We benefit from predictive navigation. Each enhancement strengthens trust. Trust encourages further data sharing.

This cycle embeds dependency. As systems become more accurate, manual alternatives feel inefficient. Why search manually when recommendations are immediate. Why compare prices when suggestions appear tailored. Efficiency masks consolidation of control.

The consolidation of data within a handful of corporations amplifies economic inequality. Startups struggle to compete without comparable datasets. Dominant platforms acquire emerging competitors to maintain advantage. Market power concentrates.

Influence also shapes perception of reality. Trending topics signal collective attention. Viral posts imply social consensus. High follower counts convey authority. Metrics create hierarchies of visibility.

These hierarchies influence career trajectories. Content creators adjust output to algorithmic preferences. Journalists craft headlines optimized for clicks. Businesses tailor messaging for platform ranking. Creative expression adapts to data feedback.

The currency of influence transforms human experience into quantifiable assets. Emotions become engagement signals. Relationships become network graphs. Preferences become targeting parameters. The richness of life is translated into analytics.

Political systems are not immune. Campaign strategies increasingly rely on data analytics. Voter segmentation enables tailored messaging. Issue framing adapts to micro-demographics. Democracy intersects with marketing science.

The ethical implications are profound. When influence is commodified, authenticity becomes secondary. Decisions are shaped by unseen calculations. Transparency lags behind capability. Citizens struggle to understand the forces acting upon them.

Data accumulation also raises privacy concerns. Breaches expose sensitive information. Misuse undermines trust. Regulations attempt to keep pace but often trail innovation. Legal frameworks strain under technological acceleration.

Yet data itself is not inherently harmful. It can improve healthcare outcomes through predictive diagnostics. It can enhance urban planning through traffic analysis. It can advance scientific research through pattern recognition. The issue lies in governance and incentive alignment.

When profit is the primary metric, influence gravitates toward revenue optimization. When public good becomes a metric, influence can be redirected. The structure of incentives determines outcomes. Data amplifies whichever goal is prioritized.

Individuals often underestimate their collective power. A single user's data may seem insignificant. Aggregated across millions, patterns emerge with remarkable precision. Collective behavior becomes a resource exploited at scale.

The metaphor of currency is apt because data fuels transactions. Advertisers purchase access to attention. Platforms monetize behavioral insights. Users pay with personal information instead of cash. The economy revolves around prediction.

This predictive economy shifts cultural norms. Privacy becomes negotiable. Sharing becomes expected. Oversharing becomes normalized. Boundaries blur between public and private life.

Influence also extends to self-perception. When algorithms suggest content aligned with past behavior, they reinforce identity patterns. Exploration narrows. Novelty decreases. Personal growth can stagnate within curated loops.

Breaking free requires awareness of how influence operates. It requires questioning recommendations. It requires diversifying information sources. It requires understanding that convenience carries hidden costs.

Regulatory frameworks can address asymmetry by demanding transparency. Data portability can empower users. Ethical design can reduce exploitative practices. Market competition can limit monopolistic dominance. Solutions exist but require collective will.

The currency of influence will only increase in value as artificial intelligence advances. Predictive models will grow more sophisticated. Behavioral insights will deepen. The capacity to steer outcomes will expand.

Whether this capacity serves human flourishing or corporate dominance depends on choices made now. Influence is not destiny. It is a tool shaped by incentive structures. Redirecting incentives can reshape outcomes.

Gold glittered visibly. Data glows invisibly in server racks. Its weight cannot be felt in the hand, but its impact is measurable in markets and minds. The wealth of the digital era is behavioral insight.

Understanding this shift is essential to navigating the afterworld. When we recognize data as currency, we better grasp why attention is contested so fiercely. We see why platforms resist transparency. We understand why influence is subtle rather than overt.

Power no longer marches in uniforms. It flows through code. It scales through networks. It operates continuously. Recognizing its mechanisms is the first step toward reclaiming balance.

The question is not whether data should exist. It is who controls it and for what purpose. Currency reflects values. If influence remains unchecked, imbalance deepens. If influence is guided ethically, possibility expands.

We live in an era where information about us is more valuable than information we possess. That inversion defines the modern hierarchy. To challenge it, we must demand accountability, transparency, and intentional design.

The currency of influence will shape the century ahead. Whether it becomes a force for collective empowerment or concentrated control depends on how consciously we engage with it. Awareness transforms passive subjects into active citizens. And active citizens can redefine the economy of attention.

CHAPTER . 4

The Collapse of Shared Reality

A functioning society depends on shared reference points. Citizens may disagree about policy, morality, or culture, but they rely on common facts to frame those disagreements. Newspapers once served as centralized arbiters of daily events. Broadcast networks delivered the same headlines into millions of homes. While bias always existed, there was overlap in informational exposure. That overlap created a baseline of shared reality.

The digital revolution dissolved that baseline. Information no longer flows through a limited number of gatekeepers. It moves through decentralized platforms optimized for personalization. Each user receives a tailored stream shaped by past behavior. The result is fragmentation.

Fragmentation does not merely diversify perspective. It isolates perception. When two individuals search for the same topic and receive different results, they inhabit parallel narratives. When news feeds prioritize divergent sources, consensus becomes elusive. Reality splinters into algorithmic silos.

The architecture of personalization accelerates this process. Algorithms learn which stories users engage with most frequently. They amplify similar content to increase retention. Over time, exposure narrows. Contrary viewpoints appear less often.

This narrowing feels comfortable. Confirmation bias is psychologically soothing. Information that aligns with prior belief reduces cognitive strain. The brain rewards coherence. The algorithm detects this reward pattern and supplies more alignment.

The effect is gradual and often unnoticed. Individuals rarely sense the shrinking perimeter of their informational world. They perceive abundance because content is constant. Yet abundance of quantity does not equal diversity of viewpoint. Repetition masquerades as consensus.

Shared reality requires friction. It requires encountering ideas that challenge assumptions. In traditional civic spaces, such friction was unavoidable. Diverse communities interacted physically. Editorial standards constrained extreme claims. Public discourse unfolded within bounded arenas.

Digital spaces remove many of these constraints. Anyone can publish content instantly. Verification often occurs after dissemination, if at all. False claims can travel rapidly before correction. Emotional resonance outpaces factual nuance.

The speed of digital communication compounds the problem. Stories circulate globally within minutes. Viral posts accumulate millions of impressions before context emerges. Corrections rarely travel as far as initial misinformation. The first impression often persists.

Algorithmic amplification prioritizes engagement over accuracy. Content that provokes strong emotion spreads more widely. Fear and outrage are particularly contagious. The design does not distinguish between truth and falsehood. It measures interaction.

As a result, misinformation ecosystems flourish. Niche communities coalesce around specific narratives. Reinforcement within these communities strengthens conviction. Skepticism toward external sources increases. Trust erodes.

Trust is the glue of shared reality. When institutions lose credibility, alternative authorities emerge. Influencers replace journalists. Anonymous accounts replace editors. Popularity replaces verification. The hierarchy of knowledge shifts.

The collapse of shared reality manifests politically. Citizens consume divergent media ecosystems aligned with ideological preferences. Debates become less about policy differences and more about factual disagreement. Compromise becomes difficult when foundational premises diverge.

It also manifests culturally. Trends rise and fall within subcultures insulated from broader awareness. Micro-communities develop distinct vocabularies and norms. While diversity enriches society, isolation fragments it. Without bridges, difference becomes division.

The phenomenon extends beyond politics. Public health crises reveal fractures in informational consensus. Scientific guidance competes with anecdotal claims. Expert opinion struggles against viral speculation. Confusion proliferates.

Technology companies attempt moderation through content policies. Yet moderation at scale is complex and contentious. Decisions about removal or labeling are interpreted through partisan lenses. Efforts to curb misinformation can be framed as censorship.

This tension exposes a structural dilemma. Platforms built for open expression now function as de facto public squares. Their design incentives prioritize engagement. Their societal role demands responsibility. The two aims often conflict.

The collapse of shared reality is not solely technological. It reflects human psychology interacting with digital architecture. We gravitate toward groups that affirm identity. We resist information that threatens belonging. Algorithms intensify these tendencies.

Echo chambers form when exposure becomes self-reinforcing. Within these chambers, narratives evolve without external correction. Shared language strengthens group cohesion. Outsiders are viewed with suspicion. The boundary between community and cult can blur.

Polarization deepens when opposing groups rarely interact constructively. Digital debate often rewards performative aggression. Nuanced discussion struggles to gain traction. Simplified narratives dominate timelines.

The economic model contributes to this dynamic. Outrage drives clicks. Clicks drive revenue. Revenue sustains platforms. The cycle incentivizes sensational framing.

In such an environment, objective truth competes with persuasive storytelling. Fact-checking organizations attempt to intervene. Educational initiatives promote media literacy. Yet structural incentives remain powerful.

The collapse of shared reality carries psychological costs. Constant exposure to conflicting claims generates fatigue. Individuals may retreat from news entirely. Apathy replaces engagement. Democratic participation weakens.

Alternatively, some double down within ideological enclaves. Certainty becomes armor against complexity. Doubt is interpreted as weakness. The capacity for dialogue diminishes.

Shared reality once depended on temporal synchronization as well. Communities watched the same broadcasts at the same time. Now information consumption is asynchronous and individualized. Collective experience fragments.

Restoring shared reality does not require uniformity of thought. It requires agreement on methods of verification. It requires trust in processes of evidence evaluation. It requires transparency in algorithmic curation.

Digital literacy becomes crucial in this landscape. Citizens must understand how feeds are constructed. They must question sources and incentives. They must diversify information intake intentionally.

Platforms can contribute by adjusting ranking systems to prioritize credible sources. Transparency about recommendation criteria can build trust. Design choices can reduce amplification of demonstrably false claims.

Policy frameworks may also play a role. Regulatory approaches can encourage accountability without stifling innovation. Balancing free expression with harm reduction remains challenging but necessary.

Community initiatives can bridge divides. Structured dialogues across ideological lines can humanize opponents. Offline engagement can soften digital caricatures. Exposure to diverse perspectives can rebuild empathy.

Ultimately, shared reality depends on collective commitment. It requires valuing truth over virality. It requires patience for complexity. It requires humility about personal certainty.

The collapse we witness is not irreversible. It is a product of systems designed without full awareness of social consequence. Systems can be redesigned. Incentives can be recalibrated.

We must first acknowledge the fracture. Pretending consensus still exists where it does not prevents meaningful repair. Recognition allows intentional response.

The neon glow of personalized feeds can obscure the extent of fragmentation. Each user feels informed. Few perceive the divergence between informational worlds. Awareness begins with comparison.

When individuals deliberately seek out varied perspectives, they expand cognitive horizons. When institutions commit to transparency, they rebuild credibility. When platforms align design with civic responsibility, they strengthen democratic resilience.

Shared reality is fragile but essential. Without it, cooperation falters. With it, disagreement becomes productive rather than destructive. The future of pluralistic society depends on restoring common ground.

The collapse of shared reality is one of the defining challenges of the digital age. It reveals how technological architecture shapes epistemology. It shows how incentives sculpt perception. It underscores the urgency of intentional reform.

If we understand how fragmentation emerged, we can begin constructing bridges. Those bridges will require technological innovation, cultural humility, and institutional accountability. They will require effort from individuals and organizations alike.

Reality need not remain fractured. But rebuilding it demands conscious participation in the informational ecosystems we inhabit. The work begins with recognizing that our feeds are not neutral windows. They are curated mirrors.

CHAPTER . 5

The Seduction of Infinite Scroll

There was a time when content had edges. A newspaper ended at the final page. A television broadcast signed off for the night. Even a bookstore closed its doors at a predictable hour. These boundaries signaled completion. They invited pause, digestion, and reflection.

Infinite scroll erased the edge. It replaced completion with continuation. Instead of turning a page, we glide downward into an endless stream. There is no bottom, no natural stopping point. The experience feels fluid, frictionless, and inexhaustible.

This design was not accidental. Engineers understood that removing stopping cues increases engagement time. Each additional second translates into measurable value. The longer a user remains within the platform, the more advertisements can be displayed. What appears seamless is strategically constructed.

The psychology behind infinite scroll is rooted in variable reward. When rewards are unpredictable, behavior intensifies. Slot machines operate on similar principles. A swipe may reveal something entertaining, shocking, affirming, or trivial. The uncertainty keeps the thumb moving.

Completion once offered satisfaction. Now continuation offers possibility. The next post might be more interesting than the last.

The next video might answer a lingering curiosity. The promise of something better hovers just below the fold. Curiosity becomes compulsion.

The absence of boundaries blurs time perception. Minutes dissolve into hours without conscious notice. Users often emerge from scrolling sessions surprised by elapsed time. The design encourages immersion without awareness. The clock becomes irrelevant.

Friction once served as a natural regulator. Turning a page required physical effort. Waiting for a program to air required patience. Infinite scroll eliminates friction. Content loads automatically. Autoplay initiates without consent. The system anticipates inertia.

In this environment, boredom becomes intolerable. Micro-moments once filled with daydreaming are now opportunities for consumption. Waiting in line transforms into scrolling time. Elevators become portals to feeds. Silence feels empty rather than restorative.

The seduction lies in ease. There is no decision required to continue. The body moves almost independently of intention. The thumb performs the ritual repeatedly. Muscle memory reinforces the pattern.

Over time, this pattern rewires expectation. The brain grows accustomed to rapid novelty. Slow tasks feel tedious by comparison. Deep reading requires greater effort. Conversations without digital interruption feel unusually demanding.

The architecture of infinite scroll prioritizes quantity over depth. Content appears in bite-sized fragments optimized for quick reaction. Complex ideas are compressed into short posts. Nuance struggles within limited character counts.

The rapid succession of stimuli fragments attention. Each item competes briefly before yielding to the next. The mind rarely lingers long enough to analyze deeply. Reflection becomes secondary to reaction.

This constant stimulation elevates baseline arousal. The nervous system adapts to frequent novelty. Calm states may feel underwhelming. Restlessness increases when external input slows. Stillness becomes unfamiliar.

Platforms track engagement meticulously. If a user pauses on a specific type of content, similar material appears more frequently. The feed becomes increasingly tailored. Personalization deepens immersion.

Immersion can be pleasurable. Laughter, inspiration, and discovery are genuine experiences. Yet pleasure intertwined with compulsion creates ambiguity. Users may enjoy scrolling while simultaneously wishing to stop. Desire conflicts with intention.

The economic model reinforces excess. Advertisers pay for impressions. Impressions require time. Time requires sustained scrolling. The structure subtly incentivizes endlessness.

Infinite scroll also alters narrative structure. Stories unfold in fragmented updates rather than cohesive arcs. News appears alongside memes and advertisements. Context dissipates within juxtaposition. Serious issues compete with trivial diversions.

This juxtaposition influences emotional processing. Tragedy may be followed immediately by humor. Outrage may be interrupted by entertainment. Emotional continuity fractures. The mind shifts gears rapidly.

Such rapid shifting can contribute to fatigue. Emotional energy disperses across countless stimuli. Depth of feeling may diminish. Empathy competes with distraction.

Children raised within infinite scroll environments may develop different attention patterns. Early exposure to constant novelty shapes neural pathways. Patience for sustained effort may decrease. Immediate gratification becomes normative.

Adults are not immune. Professionals find themselves checking feeds between tasks. Meetings are punctuated by discreet glances at screens. Productivity fragments under intermittent digital intrusion.

The seduction persists because it offers escape. Difficult emotions can be deferred through distraction. Anxiety can be numbed temporarily. Loneliness can be masked by virtual presence. Infinite scroll becomes a coping mechanism.

Yet coping through avoidance often prolongs discomfort. Underlying concerns remain unresolved. Digital immersion delays confrontation with reality. The cycle continues.

Breaking free from infinite scroll requires conscious interruption. Disabling autoplay restores friction. Setting time limits introduces boundaries. Moving devices out of reach reduces reflexive engagement.

Some platforms now include reminders encouraging breaks. These features acknowledge potential harm. However, they remain optional. Responsibility largely falls on users.

Design reform could embed natural stopping points. Feeds could end deliberately. Content could encourage reflection rather than acceleration. Such changes would require recalibrating profit models.

The broader cultural shift must also address expectation. Valuing depth over volume requires collective reorientation. Celebrating sustained focus challenges the dominance of rapid consumption.

Infinite scroll symbolizes the ethos of the digital age. Limitless availability equates to progress. More content implies more freedom. Yet without boundaries, freedom can morph into captivity.

Captivity here is subtle. It manifests as diminished agency over time. The user believes they are choosing to continue, yet design strongly influences behavior. Awareness reveals the tension.

The seduction loses some power when named. Recognizing variable reward patterns clarifies compulsion. Understanding design incentives demystifies endlessness. Knowledge empowers restraint.

Communities can support healthier habits collectively. Shared device-free gatherings normalize disconnection. Workplaces can establish communication boundaries. Families can model intentional usage.

Reclaiming time is an act of resistance in an economy built on attention extraction. Choosing to close an app interrupts the revenue cycle. Choosing to read deeply reorients cognition. Choosing silence restores equilibrium.

Infinite scroll thrives on inertia. It weakens under deliberation. When users pause to ask why they continue, momentum falters. Conscious choice interrupts automaticity.

The solution is not total rejection of digital platforms. They provide information, connection, and entertainment. The challenge lies in balancing utility with autonomy. Boundaries transform tools from tyrants into instruments.

Edges must be reintroduced intentionally. Finite sessions can replace endless drift. Scheduled engagement can replace constant availability. Reflection can follow consumption.

The seduction of infinite scroll reflects a broader tension between design and discipline. Technology amplifies human tendencies. Discipline tempers amplification. The interplay defines modern experience.

As we confront the afterworld shaped by neon glow and endless feeds, we must decide whether continuation serves us. If not, we must craft new rituals. Rituals of closure. Rituals of pause.

The bottom of the feed does not exist by design. But the bottom of our attention span is real. Protecting it requires vigilance. Reclaiming it requires intention.

Infinite scroll promises infinite novelty. It rarely delivers infinite fulfillment. Recognizing that distinction is the first step toward freedom. And freedom begins where endlessness ends.

CHAPTER . 6

Virtual Identity and the Mask Economy

Identity was once shaped primarily by geography. The community around you determined reputation, opportunity, and belonging. People knew one another through repeated physical interaction. Behavior accumulated into character over time. Masks were difficult to maintain in small, interconnected worlds.

The digital age altered that structure fundamentally. Online, identity is constructed through profiles, posts, and images. It is curated rather than merely lived. Individuals select which moments to display and which to conceal. The result is a version of self optimized for perception.

This curation is not inherently deceptive. Humans have always presented themselves differently in varied social contexts. Professional attire differs from casual clothing. Public speech differs from private conversation. The internet, however, amplifies this performative flexibility.

Profiles become stages. Metrics become applause. The audience is potentially global. Feedback is immediate and quantifiable. Validation appears in the form of likes, shares, and comments.

In this environment, identity becomes strategic. Users quickly learn which types of content generate affirmation. Certain aesthetics attract more attention. Certain opinions receive stronger reactions. The algorithm rewards patterns.

The reward system shapes self-expression. If vulnerability receives support, vulnerability increases. If outrage garners engagement, outrage intensifies. Over time, identity aligns with performance incentives. Authenticity blends with optimization.

The mask economy emerges from this dynamic. A mask is not necessarily false, but it is selective. It emphasizes specific traits while minimizing others. Online, masks can be adjusted rapidly. Profiles can be edited. Posts can be deleted. Personas can be rebranded.

This fluidity offers freedom. Individuals can explore interests without geographic constraint. Marginalized voices can find supportive communities. Creativity can flourish beyond local limitations. The digital world expands possibility.

Yet freedom carries pressure. The awareness of constant visibility encourages self-monitoring. Users may hesitate before expressing unpopular opinions. Fear of backlash shapes discourse. Social risk becomes amplified.

The permanence of digital footprints intensifies caution. A post shared impulsively can resurface years later. Screenshots preserve deleted content. Context collapses across time and audience. Identity becomes archival.

The mask economy also commodifies selfhood. Influencer culture exemplifies this shift. Personal narratives transform into brand strategies. Daily routines become content streams. Intimacy becomes monetizable.

Followers represent both community and currency. Higher follower counts attract sponsorships. Engagement metrics determine advertising rates. Personal identity intertwines with commercial viability. The line between living and marketing blurs.

This blending can distort motivation. Experiences may be evaluated for shareability rather than intrinsic value. Vacations become photo opportunities. Meals become aesthetic compositions. Moments become material.

Comparison intensifies under these conditions. Exposure to curated highlights from countless lives creates unrealistic benchmarks. Individuals compare their unfiltered reality to others' edited presentations. Insecurity often follows.

The mask economy thrives on aspiration. Platforms amplify content that inspires envy or admiration. Luxury lifestyles, physical perfection, and professional success dominate feeds. Ordinary life appears underrepresented.

The disparity between presentation and reality can generate internal conflict. Users may feel pressure to maintain an image inconsistent with their lived experience. Emotional labor increases. Authentic connection decreases.

At the same time, anonymity enables alternate masks. Some users adopt pseudonyms to express views without reputational consequence. Anonymity can encourage honesty. It can also embolden hostility.

Without face-to-face accountability, social norms shift. Harsh criticism becomes easier. Empathy may diminish. The psychological distance between sender and receiver reduces inhibition.

Virtual identity fragments across platforms. A professional profile may differ dramatically from a personal one. A gaming avatar may embody traits absent offline. Multiple selves coexist digitally.

Fragmentation can enrich self-understanding. Exploring different aspects of identity can foster growth. However, fragmentation can also create dissonance. Maintaining consistency across contexts requires effort.

The algorithm reinforces identity patterns by feeding users content aligned with prior engagement. If someone frequently interacts with fitness content, their feed fills with similar material. Identity narrows through repetition.

Narrowing influences aspiration. Users may feel compelled to conform to subcultural norms prevalent in their digital communities. Divergence risks reduced engagement. Conformity ensures visibility.

Adolescents are particularly vulnerable within the mask economy. Identity formation is already complex during developmental years. Online validation adds an additional layer of evaluation. Peer approval becomes quantified.

Self-worth may become entangled with metrics. A post that receives fewer likes can feel like personal rejection. External affirmation becomes a proxy for internal confidence. Emotional resilience is tested.

The commercialization of identity extends beyond influencers. Everyday users contribute data that platforms monetize. Personal preferences inform advertising algorithms. Self-expression becomes input for corporate analytics.

In this sense, identity is both performance and product. The more detailed the profile, the more precisely advertisements can target. Self-disclosure fuels economic engines.

Navigating this environment requires discernment. Users must differentiate between authentic sharing and algorithmic pressure. They must evaluate whether expression aligns with personal values or platform incentives.

Digital literacy includes understanding how algorithms amplify certain identities. Recognizing this influence enables conscious resistance. Individuals can choose to diversify engagement intentionally.

Boundaries are essential. Not every moment requires documentation. Not every opinion requires broadcast. Selective absence can protect psychological well-being.

Community norms can also evolve. Encouraging authenticity over perfection reduces pressure. Celebrating vulnerability over spectacle fosters connection. Supporting nuance over extremity strengthens dialogue.

Platforms bear responsibility as well. Design choices can reduce metric visibility. Emphasizing meaningful interaction over public counts may alleviate performance anxiety. Prioritizing privacy can restore autonomy.

The mask economy reflects a broader societal shift toward visibility as value. Being seen equates to existing. Silence equates to irrelevance. Challenging this equation requires cultural recalibration.

Offline relationships provide grounding. Face-to-face interaction reveals complexity beyond curated profiles. Physical presence tempers exaggeration. Shared experiences deepen trust.

Reclaiming identity from algorithmic influence involves introspection. Asking why we share clarifies intention. Evaluating emotional response to feedback reveals dependency. Conscious choice interrupts automatic performance.

Virtual identity will remain integral to modern life. The goal is not abandonment but alignment. Technology should extend self-expression without distorting it. Platforms should facilitate connection without commodifying authenticity.

The mask is not inherently harmful. It becomes problematic when it replaces the face entirely. Balance requires remembering that online persona is a fragment, not the whole.

In the neon glow of digital spaces, identity can shimmer attractively. But enduring self-worth emerges from integrated experience, not curated presentation. Protecting that integration demands awareness.

The economy of masks thrives on unconscious adaptation. It weakens when individuals prioritize integrity over engagement. It softens when communities value depth over display.

We stand at a crossroads between performance and presence. Choosing presence does not reject digital life. It reframes it. Identity can remain expansive without becoming transactional.

The challenge of the afterworld is not merely technological. It is existential. Who are we when metrics disappear. Who are we beyond the feed. Answering these questions anchors us beyond the mask.

CHAPTER . 7

Manufactured Outrage

Anger has always been a powerful social force. It can mobilize communities against injustice and demand accountability from institutions. It can expose corruption and ignite reform. But anger can also be manipulated. In the digital age, outrage has become both currency and catalyst.

Outrage spreads faster than calm analysis. Emotionally charged content captures attention immediately. It interrupts scrolling and demands reaction. The brain prioritizes perceived threats, even symbolic ones. Platforms measure that urgency as engagement.

Algorithms do not distinguish between righteous indignation and manufactured provocation. They detect intensity. Posts that generate rapid comments, shares, and reactions rise to the top of feeds. The more explosive the response, the broader the reach. This dynamic incentivizes escalation.

Content creators quickly learn what performs well. Nuanced arguments often receive modest engagement. Simplified, emotionally framed claims travel farther. Outrage becomes a strategy rather than a spontaneous response. Performance replaces deliberation.

The architecture of social media compresses complex issues into brief statements. Context collapses into headlines. Subtlety struggles in character-limited environments. Binary framing dominates discourse. Issues are presented as battles rather than conversations.

This compression amplifies misinterpretation. Statements stripped of context can appear inflammatory. Screenshots circulate without background. Audiences react to fragments rather than full narratives. Reaction precedes understanding.

Manufactured outrage thrives on partial information. A selectively edited clip can spark global backlash. A misleading headline can shape perception before correction emerges. Emotional reaction often outpaces fact-checking.

Outrage also fosters group cohesion. Shared anger reinforces in-group identity. Expressing indignation signals loyalty. Silence may be interpreted as complicity. Social pressure intensifies participation.

The dopamine response associated with social validation reinforces this cycle. Posts expressing strong emotion often receive immediate feedback. Likes and comments provide affirmation. The emotional charge deepens.

Over time, baseline sensitivity may increase. Minor disagreements can trigger disproportionate responses. The threshold for outrage lowers. Public discourse becomes volatile.

Media outlets are not immune to these incentives. Headlines crafted for maximum click-through often emphasize conflict. Conflict drives traffic. Traffic drives revenue. The economic structure rewards emotional amplification.

Political actors exploit this environment deliberately. Provocative statements attract media coverage. Polarizing rhetoric dominates news cycles. Outrage becomes a tool for visibility.

Bots and coordinated networks further distort perception. Artificial amplification can create the illusion of widespread anger. Trending topics may reflect manipulation rather than organic consensus. Distinguishing authentic sentiment from orchestrated campaigns becomes difficult.

The psychological toll of constant outrage is significant. Chronic exposure to inflammatory content elevates stress levels. The nervous system remains in heightened alert. Fatigue accumulates.

Outrage fatigue can lead to cynicism. When every issue is framed as crisis, urgency loses potency. Individuals may disengage entirely. Democratic participation suffers.

Alternatively, sustained anger can narrow empathy. Opposing viewpoints are perceived as threats rather than perspectives. Dialogue becomes confrontation. Compromise appears as betrayal.

The public square shifts from deliberative exchange to performative spectacle. Users craft statements to maximize reaction rather than insight. The loudest voices dominate. The moderate middle recedes.

Manufactured outrage also obscures genuine injustice. When trivial controversies receive equal attention as serious issues, proportionality disappears. Moral energy disperses across endless disputes. Focus fragments.

The speed of digital reaction leaves little room for reflection. Immediate responses often precede full comprehension. Retractions rarely receive equal visibility. Reputations can be damaged permanently by viral outrage.

Cancel culture debates illustrate this tension. Accountability for harmful behavior is necessary. Yet disproportionate punishment driven by viral momentum can undermine fairness. Context matters, but context rarely trends.

Platforms attempt moderation through policy enforcement. However, defining acceptable speech across diverse cultures is complex. Decisions often provoke further outrage. Moderation itself becomes controversial.

Education in emotional regulation becomes increasingly important. Recognizing triggers and pausing before reacting interrupts escalation. Deliberate delay reduces impulsive amplification.

Journalistic standards can counteract outrage cycles. Emphasizing verification and context restores credibility. Highlighting constructive solutions rather than solely conflict can shift tone.

Individuals can curate feeds intentionally. Muting inflammatory accounts reduces exposure. Diversifying sources broadens perspective. Choosing depth over immediacy tempers reactivity.

Community norms also influence discourse. Encouraging respectful disagreement fosters resilience. Valuing empathy over victory transforms debate into dialogue. Cultural expectations shape online behavior.

The incentive structure of platforms remains central. If engagement metrics prioritize emotional intensity, outrage will persist. Redesigning algorithms to reward constructive interaction could mitigate volatility.

Transparency about amplification criteria can empower users. Understanding why certain content appears fosters critical awareness. Opaque systems breed suspicion.

Outrage itself is not inherently negative. It signals perceived violation of values. The challenge lies in distinguishing authentic moral response from manipulated provocation. Awareness sharpens discernment.

Slowing down digital interaction can reduce escalation. Introducing friction before sharing inflammatory content encourages reconsideration. Small design adjustments can alter large-scale patterns.

Offline relationships counterbalance digital volatility. Face-to-face conversation humanizes opponents. Nuance emerges more naturally. Empathy strengthens through shared presence.

The culture of instant reaction erodes patience. Reclaiming patience restores perspective. Complex issues require time to understand fully. Quick judgments rarely capture depth.

Manufactured outrage thrives in environments of uncertainty. Clear communication from institutions reduces rumor proliferation. Transparency builds trust.

Ultimately, outrage becomes harmful when it eclipses constructive action. Anger should catalyze solutions, not perpetuate division. Without direction, it dissipates into noise.

The neon glow of viral posts can make conflict feel constant. Yet outside the feed, many communities function with cooperation and compromise. Recognizing this distinction tempers perception.

The afterworld we inhabit amplifies extremes because extremes are profitable. Challenging this model requires collective intention. Individuals, institutions, and platforms must align incentives with well-being.

We must learn to differentiate signal from spectacle. Not every trending topic warrants emotional investment. Selective engagement preserves mental clarity.

Outrage will always exist in human society. The question is whether it is harnessed for justice or exploited for profit. Conscious participation determines outcome.

The manufactured nature of much digital anger becomes evident when examined closely. Patterns repeat predictably. Provocations follow familiar scripts. Recognizing repetition reduces susceptibility.

Cultivating curiosity instead of immediate condemnation shifts tone. Asking questions invites understanding. Listening precedes judgment. Dialogue replaces declaration.

In an environment engineered for reaction, restraint becomes radical. Pausing before sharing disrupts the algorithmic cascade. Choosing empathy interrupts polarization.

Manufactured outrage may dominate timelines, but it need not dominate consciousness. Awareness restores choice. Choice restores agency.

If we learn to navigate anger without being consumed by it, we weaken one of the most potent forces in the digital economy. And in doing so, we create space for more thoughtful, humane discourse in the afterworld we now inhabit.

CHAPTER . 8

The Illusion of Connection

Never in human history have so many people been reachable at once. A message can cross continents in seconds. A video call can collapse oceans into screens. Social platforms allow individuals to maintain contact with hundreds, even thousands, of acquaintances simultaneously. By every measurable metric, we are more connected than any generation before us.

And yet, loneliness is rising.

The paradox defines the digital age. Visibility has increased, but intimacy has not necessarily followed. We witness fragments of each other's lives continuously, but witnessing is not the same as knowing. Awareness of activity can masquerade as relationship. Passive consumption can feel like participation.

Connection requires reciprocity, vulnerability, and sustained attention. Digital platforms often provide only partial substitutes. A "like" signals acknowledgment but not depth. A comment may spark interaction but rarely fosters sustained dialogue. Exchanges are brief and easily displaced by the next notification.

The architecture of social media encourages breadth over depth. Users accumulate friends and followers rapidly. Networks expand horizontally rather than vertically. The emphasis rests on reach rather than resonance.

Quantification reinforces this emphasis. Follower counts become visible indicators of social capital. The number of connections may feel like a measure of belonging. Yet numerical abundance does not guarantee emotional support.

Online communication also filters emotional nuance. Text lacks tone unless carefully constructed. Emojis attempt to compensate, but subtleties can be lost. Misunderstandings arise more easily without nonverbal cues.

Video platforms restore some facial expression, yet even they mediate presence. Eye contact becomes approximate rather than direct. Delays interrupt rhythm. Shared physical space, with its sensory richness, cannot be fully replicated.

The illusion of connection intensifies through constant updates. Seeing frequent posts from acquaintances creates a sense of proximity. We feel informed about milestones and struggles. Yet information does not equal involvement.

True intimacy requires shared experience and mutual vulnerability. Scrolling through curated highlights rarely invites either. Individuals often present polished versions of their lives. Struggles may be minimized or stylized.

This curation can create comparative distance. Observing others' celebrations without context may deepen feelings of exclusion. The highlight reel effect distorts perception of communal well-being. We assume others are thriving more than they are.

Group chats offer another form of digital connection. They can sustain friendships across distance. They can coordinate logistics efficiently. Yet even here, conversation competes with distraction. Threads splinter quickly.

Physical presence engages more senses than digital interaction. Shared meals, walks, and silence cultivate bonding through embodied experience. Touch, posture, and environment contribute to emotional attunement. Screens flatten these dimensions.

The speed of digital communication can also reduce intentionality. Quick replies replace thoughtful letters. Immediate responses become expected. The patience once required for correspondence fostered reflection.

Hyper-connectivity blurs boundaries between social spheres. Work messages intrude on family time. Social notifications interrupt professional focus. Constant availability erodes separation necessary for restorative connection.

For some, online communities provide crucial support unavailable locally. Marginalized individuals can find solidarity. Niche interests can connect globally. These benefits are genuine and transformative.

However, reliance solely on digital interaction may limit relational depth. Without physical co-presence, trust can remain fragile. Online friendships may struggle to withstand conflict. Disagreements escalate quickly without embodied empathy.

The illusion of connection also manifests in parasocial relationships. Audiences feel intimate familiarity with influencers or public figures. Daily exposure to content fosters perceived closeness. Yet the relationship is unidirectional.

Parasocial bonds can provide comfort, but they cannot replace reciprocal friendship. Emotional investment flows one way. The illusion of familiarity may substitute for real-life interaction.

The convenience of digital communication can discourage local engagement. Instead of meeting neighbors, individuals may retreat into online spaces. Geographic community weakens. Social isolation paradoxically increases.

Loneliness is not merely absence of contact. It is absence of meaningful connection. One can have hundreds of digital interactions and still feel unseen. Superficial exchange does not satisfy deeper relational needs.

The design of platforms emphasizes stimulation rather than contemplation. Conversations are interspersed with advertisements and viral content. Depth competes with distraction. Sustained dialogue requires intentional effort.

Notifications fragment attention during in-person interactions. Checking a phone mid-conversation signals divided focus. Even subtle glances can erode presence. The person physically present may feel secondary to the device.

Children observe these patterns. They internalize norms about divided attention. Modeling intentional presence becomes essential. Without it, the illusion of connection replaces authentic engagement.

Rebuilding intimacy requires deliberate boundaries. Device-free meals restore eye contact. Scheduled visits replace endless texting. Physical gatherings reinforce communal bonds.

Technology can support rather than supplant connection when used intentionally. Video calls can maintain relationships across distance when complemented by periodic in-person meetings. Messaging can coordinate shared experiences rather than replace them.

Mindful communication enhances depth. Asking open-ended questions encourages dialogue. Listening without multitasking signals respect. Slowing conversation counters digital haste.

Community spaces play a vital role. Libraries, parks, and local events create opportunities for spontaneous interaction. Shared physical environments cultivate belonging beyond algorithms.

Workplaces can foster connection by limiting after-hours digital intrusion. Clear boundaries preserve personal time. Respecting disconnection enhances well-being.

Educational institutions can teach relational skills alongside digital literacy. Empathy, conflict resolution, and active listening require practice. Screen-based interaction should not eclipse these competencies.

The illusion of connection thrives when convenience substitutes for commitment. Genuine connection requires effort. It requires vulnerability and time. These qualities resist optimization.

Reclaiming depth involves reevaluating priorities. Allocating time for uninterrupted conversation signals value. Choosing fewer, deeper relationships over many superficial ones enhances fulfillment.

The neon glow of devices can create ambient companionship. The presence of a feed may reduce feelings of solitude temporarily. Yet digital noise rarely satisfies emotional hunger.

Silence in the presence of another person differs from scrolling alone. Shared quiet can be comforting. Digital silence often feels empty because it lacks embodied context.

The afterworld we inhabit tempts us with perpetual contact. But contact is not connection. Awareness of this distinction empowers intentional choice.

Loneliness cannot be solved solely through increased bandwidth. It requires authentic encounter. It requires mutual recognition.

Platforms could redesign metrics to emphasize meaningful interaction. Prioritizing sustained exchanges over passive consumption may shift culture. Encouraging smaller, private groups could enhance intimacy.

Ultimately, connection is a human need that predates technology. Tools should serve that need rather than distort it. When we treat devices as bridges rather than destinations, balance emerges.

The illusion dissolves when we examine experience honestly. Are we nourished by our interactions or merely occupied. Do we feel known or simply visible. These questions guide recalibration.

Hyper-connectivity has expanded possibility. It has not eliminated the need for embodied community. The challenge lies in integrating digital convenience with human depth.

If we reclaim intentional presence, the illusion loses power. We can use technology to enhance relationships without mistaking it for relationship itself. In doing so, we step closer to restoring the intimacy that endless connectivity promised but did not fully deliver.

CHAPTER . 9

Dopamine, Design, and Dependency

The human brain evolved for survival, not for infinite stimulation. It developed reward systems to reinforce behaviors necessary for life. Finding food, forming bonds, and achieving goals triggered chemical responses that encouraged repetition. Dopamine played a central role in this process. It signaled anticipation and motivated pursuit.

In the digital age, this ancient circuitry encounters unprecedented intensity. Platforms are engineered to activate reward pathways repeatedly. Notifications, likes, and new content function as micro-rewards. Each signal promises something socially or intellectually valuable. Anticipation becomes habitual.

Dopamine does not create pleasure directly. It drives seeking behavior. It fuels the desire to check, to refresh, to explore. When the brain anticipates a potential reward, dopamine levels rise. The uncertainty of outcome strengthens this effect.

Variable rewards are particularly powerful. If every action produced the same predictable response, interest would decline. Instead, digital platforms deliver inconsistent reinforcement. Some posts receive high engagement, others minimal. Some refreshes reveal captivating content, others mundane.

This unpredictability sustains engagement. The brain continues seeking because the next reward might be significant. The mechanism mirrors gambling psychology. Each swipe carries possibility.

Designers understand these principles deeply. Behavioral science informs interface decisions. Color schemes, sound cues, and animation timing are tested meticulously. Small adjustments can increase user retention measurably.

Red notification badges signal urgency. Vibrations provide tactile prompts. Infinite scroll eliminates stopping cues. Autoplay removes decision friction. Each feature reduces the effort required to continue.

Over time, repeated stimulation recalibrates baseline dopamine response. The brain adapts to frequent novelty. Ordinary activities may feel less stimulating by comparison. Patience for slower rewards diminishes.

This recalibration influences attention span. Tasks requiring sustained focus compete with rapid digital feedback. Reading long-form material may feel taxing. Deep work becomes more challenging.

Dependency develops gradually. Checking a phone becomes reflexive. Moments of boredom trigger automatic scrolling. The absence of stimulation can produce mild discomfort.

That discomfort resembles withdrawal. When habitual users attempt digital detox, they may experience restlessness or irritability. The brain anticipates rewards that no longer arrive. The seeking impulse persists.

Dependency is not identical to addiction, but the overlap is notable. Compulsive use despite negative consequences signals loss of control. Sleep may suffer due to late-night scrolling. Productivity may decline due to frequent interruptions.

Social validation intensifies the reward loop. Positive feedback on posts reinforces behavior. Anticipation of response motivates continued sharing. The platform becomes a stage for intermittent affirmation.

Negative feedback can also trigger dopamine cycles. Controversial posts may generate heightened engagement. Emotional intensity increases arousal. The brain interprets heightened arousal as significance.

Notifications fragment attention throughout the day. Each alert invites a shift in focus. Even brief interruptions disrupt cognitive flow. Returning to deep concentration requires time.

This constant toggling strains mental resources. Cognitive switching incurs cost. The brain must reorient repeatedly. Fatigue accumulates.

Sleep patterns are particularly vulnerable. Blue light exposure delays melatonin production. Late-night engagement stimulates rather than calms. Devices placed beside beds invite nocturnal checking.

Chronic sleep disruption affects mood and cognition. Irritability increases. Decision-making deteriorates. Emotional regulation weakens.

Adolescents are especially sensitive to dopamine-driven design. Developing brains exhibit heightened reward sensitivity. Social approval carries amplified significance. Online validation can feel intensely consequential.

Early immersion in digital reward cycles may shape long-term habits. Expectations of constant stimulation can form before self-regulation skills mature. Parental guidance becomes critical.

Adults are not immune to these mechanisms. Professionals may rationalize constant connectivity as productivity. Yet frequent checking often masks compulsion. The line between necessity and habit blurs.

Recognizing the neurochemical basis of engagement empowers change. Understanding that impulses are engineered reduces self-blame. It shifts focus toward design awareness.

Practical interventions can recalibrate reward systems. Disabling nonessential notifications reduces triggers. Setting specific times for checking platforms introduces structure. Moving devices out of immediate reach limits reflex.

Mindfulness practices strengthen awareness of impulse. Observing the urge to check without acting interrupts automatic behavior. Pausing creates space for choice.

Engaging in activities with slower reward cycles restores balance. Reading books, practicing instruments, or exercising outdoors provide sustained fulfillment. These activities recalibrate dopamine response gradually.

Physical movement plays a role in regulation. Exercise releases neurotransmitters that stabilize mood. Natural environments reduce stress markers. Offline experiences anchor attention.

Design reform at the platform level could mitigate dependency. Introducing natural stopping cues would reduce endless engagement. Transparent metrics about time spent could increase awareness.

Some devices now include screen-time tracking features. These tools quantify usage patterns. Data about personal habits can motivate adjustment.

Community accountability enhances change. Shared commitments to device-free time reinforce discipline. Collective norms reduce social pressure to respond instantly.

Educational curricula can include neuroscience literacy. Teaching students how dopamine functions fosters informed digital behavior. Awareness reduces vulnerability.

The goal is not demonizing dopamine. It is essential for motivation and joy. The challenge lies in preventing artificial overstimulation from distorting natural reward balance.

Technology can enhance learning and connection when used intentionally. The problem arises when design prioritizes compulsion over well-being. Incentive structures shape outcomes.

Policy discussions increasingly consider platform responsibility. Ethical standards for persuasive design may emerge. Transparency about behavioral research could become normative.

Ultimately, dependency reflects interaction between biology and architecture. We cannot alter human neurochemistry easily, but we can modify digital environments. Aligning design with psychological health is possible.

Reclaiming autonomy begins with awareness of impulse. Recognizing that the urge to scroll is partly engineered reframes experience. Choice becomes visible again.

Small changes accumulate. Reducing screen time incrementally can restore focus. Establishing device-free rituals can rebuild attention span. Celebrating offline accomplishments reinforces alternative rewards.

The psychological afterworld we inhabit amplifies seeking behavior. But seeking can be redirected toward meaningful pursuits. Dopamine can motivate creativity, learning, and connection.

Dependency weakens when intention strengthens. Awareness disrupts automaticity. Deliberate engagement replaces reflex.

The brain remains adaptable throughout life. Neural pathways adjust to new habits. With sustained effort, overstimulation can subside.

The design of our digital environment influences our mental landscape profoundly. Recognizing this interplay is foundational to reclaiming balance. We are not powerless within the system.

Dopamine responds to what we repeatedly pursue. If we pursue depth rather than distraction, reward circuits recalibrate. If we pursue presence rather than perpetual novelty, satisfaction deepens.

The neon glow of constant notification need not dominate consciousness. Understanding the chemistry behind compulsion reveals leverage points. And with leverage comes the possibility of freedom.

CHAPTER . 10

Anxiety in the Age of Comparison

Comparison is a natural human behavior. For centuries, people compared themselves to neighbors, colleagues, and community members within a limited social radius. These comparisons occurred within shared environments and similar constraints. The scale was human. The reference points were local.

The digital age expanded the comparison field exponentially. Now individuals compare themselves to curated highlights from millions of lives. Social platforms present continuous streams of success, beauty, productivity, travel, and achievement. The feed becomes a global exhibition. Every scroll introduces a new benchmark.

This expansion distorts perception. We compare our private struggles to others' public victories. We compare unedited realities to carefully selected moments. Most posts represent peak experiences, not ordinary days. Yet the brain processes them as typical.

Exposure to curated content can subtly alter self-evaluation. When surrounded by images of exceptional performance or appearance, individuals may internalize unrealistic standards. The baseline for adequacy shifts upward. Ordinary life begins to feel insufficient.

This phenomenon intensifies during adolescence and early adulthood, periods already sensitive to identity formation. Peer approval carries heightened importance. Visual presentation becomes central. Platforms that emphasize imagery amplify appearance-based comparison.

Anxiety often emerges from perceived gaps between self and ideal. The larger the gap, the stronger the discomfort. Digital environments widen perceived gaps by continuously showcasing extraordinary outcomes. Success stories dominate visibility.

Algorithmic curation compounds the effect. Engagement-driven systems prioritize content that generates strong reactions. Aspirational and sensational posts often outperform mundane ones. As a result, feeds skew toward extremes.

This skew creates a distorted statistical reality. Users may believe that most people are achieving remarkable milestones regularly. In truth, platforms simply amplify outliers. The exceptional becomes commonplace in perception.

Comparison also extends to productivity. Professionals view others' accomplishments online and infer higher output standards. Entrepreneurs showcase milestones and funding announcements. Creators display rapid growth. Observers may feel behind.

The perception of falling behind can trigger chronic stress. Even when personal progress is steady, exposure to constant advancement elsewhere may undermine confidence. The brain interprets relative position as status. Status influences emotional well-being.

Anxiety in the age of comparison is reinforced by quantification. Likes, followers, and shares serve as visible indicators of social value. These metrics transform subjective qualities into numerical rankings. Visibility becomes measurable.

When self-worth becomes entangled with metrics, fluctuations in engagement affect mood. A post receiving less interaction than expected can generate disappointment. Positive feedback produces temporary elevation. Emotional stability becomes externally regulated.

The comparison environment does not operate uniformly. Certain demographics face amplified pressures. Body image comparisons are especially prevalent in visually driven platforms. Edited photos and filters can reshape norms.

Filters and editing tools blur the boundary between reality and enhancement. When enhancements become widespread, unaltered images may seem deficient. Standards evolve rapidly. The definition of normal shifts.

Anxiety is not solely about appearance. It also relates to lifestyle comparison. Vacation photos suggest constant leisure. Career announcements imply seamless advancement. Relationship posts highlight milestones. Ordinary experiences feel less noteworthy.

The continuous exposure to highlight reels can distort time perception. Individuals may feel stagnant even when making steady progress. The absence of visible comparison within local communities intensifies reliance on digital benchmarks.

Social validation mechanisms further reinforce anxiety cycles. Seeking approval can temporarily relieve insecurity. However, reliance on external affirmation creates dependency. Emotional equilibrium becomes contingent on response.

The unpredictability of feedback sustains attention. Anticipation of approval generates dopamine-driven motivation. Lack of response may trigger rumination. The mind replays interactions searching for signals of acceptance.

Chronic comparison can contribute to decreased self-esteem. Repeated exposure to superior portrayals may internalize inferiority beliefs. These beliefs influence decision-making and risk-taking. Opportunities may be avoided due to fear of inadequacy.

Anxiety also affects sleep and concentration. Persistent social evaluation increases cognitive load. The mind monitors performance across platforms. Rest becomes difficult when self-presentation remains active.

Breaking comparison cycles requires intentional curation. Diversifying feeds reduces repetitive exposure to similar content. Following accounts that promote authenticity can rebalance perception. Limiting time spent scrolling decreases cumulative impact.

Practicing gratitude counters comparative focus. Reflecting on personal progress shifts attention inward. Recognizing growth over time recalibrates standards. Internal metrics replace external benchmarks.

Community engagement offline strengthens grounded perspective. In-person interaction reveals shared vulnerabilities. Physical presence humanizes others beyond curated imagery. Real conversations expose complexity.

Educational efforts can help individuals interpret digital content critically. Understanding editing tools and algorithmic selection reduces misinterpretation. Media literacy fosters resilience against distorted standards.

Design improvements could reduce comparative pressure. Platforms might limit visible engagement metrics or allow users to hide counts. Emphasizing content quality over popularity could shift focus.

The culture surrounding social media also plays a role. Encouraging transparency about struggles balances highlight narratives. Sharing challenges alongside achievements creates realism.

Comparison is not inherently harmful. It can inspire growth when grounded in attainable reference points. Observing others' accomplishments can motivate improvement. Healthy comparison requires context and proportionality.

The problem arises when comparison becomes continuous and global. When exposure exceeds cognitive capacity to process fairly, anxiety increases. Unlimited access to curated lives intensifies pressure.

The afterworld shaped by digital systems amplifies visibility without necessarily enhancing connection. Increased visibility multiplies reference points. Without boundaries, the mind struggles to maintain equilibrium.

Reclaiming psychological stability involves limiting exposure and redefining success. Personal goals should guide evaluation rather than algorithmic trends. Self-defined progress fosters resilience.

The age of comparison challenges individuals to cultivate internal validation. Confidence rooted in values rather than metrics withstands fluctuations. Self-awareness becomes protective.

Anxiety decreases when attention shifts from others' highlight reels to lived experience. Engaging in meaningful activities reduces time available for comparison. Depth replaces distraction.

The neon streams of perfection are constructed illusions, not comprehensive portraits. Recognizing their selectivity reduces their authority. When we understand curation, we regain perspective.

Comparison will remain part of human cognition. But its scale can be managed. By designing environments and habits that emphasize authenticity, we can reduce unnecessary stress.

In the psychological afterworld, anxiety often signals misaligned exposure rather than personal deficiency. Adjusting informational diets can restore balance. Attention, when directed wisely, becomes stabilizing.

The solution lies not in withdrawing from digital life entirely but in engaging it deliberately. Limiting comparison-driven consumption strengthens mental health. Cultivating internal standards restores confidence.

Anxiety in the age of comparison is a systemic phenomenon amplified by design. Recognizing this context alleviates self-judgment. And with understanding comes the possibility of relief.

CHAPTER . 11

The Loneliness Paradox

We live in the most connected era in human history. Messages travel instantly across continents. Video calls collapse distance into pixels. Social platforms allow continuous interaction with vast networks of people. From a technological standpoint, isolation should be nearly impossible.

Yet loneliness has become one of the defining conditions of modern life.

The paradox is not about access to communication. It is about the quality of that communication. Connection in the digital age is abundant in quantity but often thin in depth. We can reach many people, but reaching is not the same as relating.

Loneliness is not simply the absence of others. It is the absence of meaningful belonging. A person can be surrounded by notifications and still feel unseen. A feed can be full of voices and still feel empty. The sensation persists even amid constant activity.

Digital platforms simulate proximity. We see updates from friends, colleagues, and acquaintances throughout the day. We witness birthdays, achievements, and personal milestones. This creates the impression of shared experience. But observation is not participation.

The human nervous system evolved for embodied interaction. Facial expressions, tone of voice, physical touch, and synchronized presence regulate emotional states. These cues signal safety and trust. Screens reduce these signals to limited visual and auditory fragments.

Without embodied feedback, subtle misunderstandings multiply. Text messages lack nuance unless carefully composed. Delays in response can feel like rejection. Typing indicators create anticipation. The absence of immediate reply can generate anxiety.

The structure of digital communication also favors brevity. Short messages replace extended conversation. Emojis compress emotion into symbols. Nuance competes with speed. Over time, depth may diminish.

Social media increases exposure but often reduces intimacy. We may know what many people are doing, yet rarely engage in extended dialogue with them. Passive consumption replaces active exchange. We scroll instead of speak.

The brain interprets meaningful interaction as bonding. Shared laughter, sustained conversation, and mutual vulnerability strengthen social connections. These experiences trigger neurochemical responses associated with trust. Digital interactions may stimulate engagement but often lack full bonding cues.

The illusion of closeness can mask relational gaps. Seeing someone's daily posts creates familiarity. Familiarity feels like friendship. However, repeated exposure without reciprocal depth does not necessarily build support.

This illusion can become especially powerful during periods of life transition. Moving to a new city, starting a remote job, or shifting social circles may reduce physical interaction. Digital platforms may fill the gap superficially. Yet substitute interaction may not satisfy deeper needs.

Remote work environments illustrate this shift. Virtual meetings provide professional coordination. Yet informal hallway conversations disappear. Casual social bonding declines. Organizational culture becomes harder to cultivate.

The loneliness paradox is intensified by algorithmic prioritization. Content that generates engagement is amplified. Emotionally charged posts often outperform neutral ones. As feeds become optimized for attention, they may prioritize stimulation over connection.

This stimulation can crowd out slower forms of interaction. Thoughtful discussion requires time and patience. Platforms designed for rapid consumption may not encourage reflective dialogue. The pace of exchange influences relational quality.

Loneliness can also increase when comparison dominates interaction. Constant exposure to others' curated highlights may generate feelings of exclusion. Even within active networks, individuals may feel peripheral. Visibility does not guarantee inclusion.

The paradox deepens because digital platforms promise community. Groups form around shared interests. Support networks emerge for niche identities. These communities can be profoundly meaningful.

However, reliance exclusively on online communities may limit embodied connection. Humans benefit from shared physical environments. Common spaces foster spontaneous encounters. These interactions strengthen trust in ways that screens cannot fully replicate.

Loneliness is associated with various psychological and physiological effects. Prolonged isolation can elevate stress responses. It can influence sleep patterns and mood regulation. Social connection functions as a stabilizing factor for well-being.

The constant availability of communication tools may paradoxically reduce deliberate outreach. Because contact is always possible, it may feel less urgent. Intentional scheduling of in-person meetings becomes less frequent. Convenience replaces commitment.

Notifications provide intermittent reassurance. A message can temporarily alleviate isolation. Yet the relief may be brief. Once the interaction ends, the underlying sense of disconnection can return.

The structure of online interaction can fragment attention even during conversation. Multitasking during calls reduces depth. Checking other applications divides presence. Divided attention limits emotional attunement.

Children and adolescents developing within this environment may experience relational shifts. Play once required shared physical space. Now digital games and platforms can substitute for local interaction. While these tools can connect peers, they may also reduce face-to-face time.

Community institutions such as schools, clubs, and local organizations play a vital role in countering loneliness. They create repeated in-person interactions. Repetition builds familiarity. Familiarity fosters trust.

Design changes could encourage healthier engagement patterns. Platforms might emphasize smaller group interactions rather than large broadcast networks. Tools could prioritize conversation threads over passive feeds. Features could highlight meaningful exchanges rather than raw engagement counts.

Individual practices also matter. Setting aside device-free periods creates space for reflection and outreach. Joining local activities strengthens social bonds. Prioritizing regular in-person gatherings reinforces belonging.

Loneliness can persist even among those with extensive digital networks. The number of contacts does not determine emotional support. Quality relationships require time and vulnerability. These qualities resist acceleration.

The paradox reveals a central tension of the digital age. Technology expands the potential for connection while simultaneously enabling shallow engagement. It offers tools for intimacy but also incentives for distraction.

Addressing loneliness requires intentional effort at multiple levels. Individuals must cultivate presence. Communities must encourage inclusion. Platforms must consider relational outcomes alongside engagement metrics.

The neon glow of screens can simulate companionship, but human beings require more than simulation. We need shared environments. We need sustained dialogue. We need mutual recognition beyond metrics.

Rebuilding connection may involve rediscovering older rhythms. Shared meals, collaborative projects, and neighborhood events anchor social life. Consistent physical interaction strengthens resilience.

The loneliness paradox does not imply that technology is inherently isolating. It means that without conscious design and intentional use, it can inadvertently reduce depth. Awareness transforms potential harm into opportunity.

When we recognize that quantity of interaction does not equal quality, we can recalibrate priorities. When we choose presence over performance, relationships deepen. When we invest time offline, belonging strengthens.

The afterworld does not have to remain emotionally distant. It can become a space where digital tools complement, rather than replace, human contact. The key lies in balancing reach with resonance.

Loneliness diminishes when individuals feel seen, heard, and valued in sustained ways. That requires attention that cannot be automated. It requires empathy that cannot be quantified.

In understanding this paradox, we take another step toward reclaiming humanity within technological systems. The tools themselves are not the enemy. Neglecting relational depth is the risk. By re-centering connection on embodied experience and intentional presence, we begin to resolve the contradiction at the heart of our hyperconnected world.

CHAPTER . 12

Digital Tribalism

Human beings evolved in small groups where belonging meant survival. Loyalty to the tribe ensured protection, shared resources, and collective defense. Our brains developed strong mechanisms for in-group identification and out-group vigilance. These mechanisms were adaptive in environments where threats were immediate and physical.

In the digital age, tribal instincts operate within vast, algorithmically amplified networks. Instead of villages, we inhabit global platforms. Instead of a few neighboring groups, we encounter countless identity-based communities. The scale has expanded, but the psychology remains ancient.

Algorithms intensify tribal dynamics by organizing content around engagement. Posts that align with a user's prior interactions are prioritized. This reinforcement strengthens existing beliefs. Over time, exposure narrows within ideological boundaries.

Within these digital tribes, shared narratives create cohesion. Members validate one another's perspectives. Agreement signals loyalty. Dissent can signal betrayal. Social rewards accumulate around conformity.

This dynamic encourages polarization. When opposing groups rarely encounter nuanced representation of each other, stereotypes solidify. Algorithms often present simplified or extreme examples of out-groups because such content generates strong reactions. Emotional intensity drives visibility.

Digital environments reduce friction between tribes. In physical communities, daily interaction with diverse neighbors can moderate extreme views. Online spaces allow selective exposure. Individuals can filter out dissenting perspectives entirely.

Echo chambers form when information flows predominantly in one direction. Inside these chambers, narratives reinforce one another without external correction. Repetition increases perceived credibility. Familiarity becomes mistaken for truth.

Tribal identity online is frequently tied to ideology, culture, or lifestyle. Visual symbols, hashtags, and shared language signal affiliation. These markers foster belonging. They also delineate boundaries.

Boundaries become sharper when engagement rewards confrontation. Outrage toward opposing tribes often receives substantial attention. Conflict generates comments and shares. Algorithms amplify what sustains interaction.

The cycle deepens when users receive affirmation from their own tribe for criticizing others. Social validation reinforces divisive rhetoric. Group cohesion strengthens internally while external hostility increases.

Digital tribalism is not limited to politics. It appears in fandoms, professional communities, wellness movements, and lifestyle groups. Any identity category can become a tribe when reinforced by algorithmic sorting.

These tribes provide genuine support. They connect individuals across geographic distance. They enable collaboration and shared purpose. For marginalized groups, digital communities can offer safety and recognition unavailable locally.

However, problems arise when tribal loyalty overrides critical thinking. When group consensus replaces independent evaluation, misinformation spreads more easily. Within closed networks, false claims may circulate unchallenged.

The speed of digital communication accelerates tribal alignment. Viral posts quickly define group positions. Members feel pressure to respond promptly to maintain alignment. Silence may be interpreted as disapproval.

Tribal dynamics also shape emotional tone. Anger toward opposing groups often signals solidarity. Humor can reinforce internal bonds. Shared memes become cultural shorthand.

The algorithmic structure contributes by clustering users with similar engagement histories. Recommendation systems group individuals based on behavior patterns. This clustering increases homogeneity of exposure.

Homogeneity strengthens internal cohesion but reduces cross-tribal understanding. Without exposure to diverse perspectives, empathy diminishes. Opponents become abstract categories rather than individuals.

In physical communities, shared spaces such as schools, workplaces, and public events foster incidental contact between diverse groups. Online environments lack these natural intersections unless intentionally designed. Separation becomes easier than dialogue.

Digital tribalism can escalate rapidly during crises. Conflicts, elections, and social controversies intensify group identification. Platforms amplify emotionally charged content. The feedback loop accelerates polarization.

This polarization has measurable societal consequences. Trust in institutions declines when narratives diverge significantly. Consensus about basic facts erodes. Cooperation across differences becomes more difficult.

Tribal identity can also affect perception of credibility. Information is evaluated based on the perceived affiliation of the source. Content aligned with group identity is trusted more readily. Out-group sources are dismissed regardless of evidence.

Such filtering simplifies cognitive processing but reduces accuracy. Identity-based evaluation bypasses independent verification. The tribe becomes the primary reference point.

Breaking out of digital tribalism requires deliberate effort. Exposure to diverse viewpoints can broaden understanding. Engaging respectfully with opposing perspectives can reduce hostility. Curiosity weakens rigid boundaries.

Platforms could design features that encourage cross-cutting dialogue. Recommendation systems might include diverse content intentionally. Transparency about clustering criteria could increase awareness of exposure limits.

Education in critical thinking is essential. Understanding cognitive biases helps individuals recognize tribal tendencies. Awareness of in-group favoritism reduces unconscious alignment.

Offline interaction further mitigates polarization. Face-to-face conversation humanizes disagreements. Physical presence introduces nuance. Emotional cues soften rigid positions.

Community initiatives that bring together diverse groups can strengthen social cohesion. Shared goals, such as local projects or volunteer efforts, create common ground. Cooperative action bridges tribal divides.

Digital tribalism thrives in environments optimized for engagement. Conflict is engaging. Agreement within tribes is reinforcing. Systems designed around attention inadvertently magnify division.

Recognizing this structural influence reduces personal blame. Individuals are not solely responsible for polarization. They operate within incentive-driven ecosystems. Awareness shifts focus from moral judgment to design reform.

The neon infrastructure of online platforms can fragment society if left unchecked. Yet technology can also facilitate cross-tribal understanding. The same tools that divide can connect when guided by different principles.

Intentional consumption habits help counter tribal narrowing. Following sources across perspectives encourages balance. Pausing before sharing emotionally charged content reduces amplification.

Digital tribalism is rooted in evolutionary psychology interacting with modern architecture. Our desire for belonging is natural. The challenge lies in preventing algorithms from exploiting that desire to intensify division.

Healthy communities require both identity and openness. Belonging does not require hostility toward outsiders. Shared values can coexist with respectful disagreement.

The afterworld need not remain fragmented. By redesigning incentives and cultivating conscious engagement, digital spaces can support pluralism rather than polarization. Cross-tribal dialogue strengthens resilience.

Tribal instincts will persist because they are part of human nature. But their expression can be shaped. Technology can either harden boundaries or build bridges. The direction depends on design and intention.

In recognizing digital tribalism, we confront one of the most powerful forces shaping contemporary discourse. Awareness allows intervention. And intervention restores the possibility of cooperation across difference.

CHAPTER . 13

The Performance of Self

In earlier eras, identity developed slowly within stable social circles. Reputation formed over time through repeated interactions. People knew one another in multiple contexts—work, family, neighborhood, and community. Character was observed across situations, not constructed for a single audience.

The digital environment compresses identity into a performance space. Every post becomes a stage. Every profile functions as a public portfolio. Individuals decide what to reveal, what to emphasize, and what to conceal. The self becomes curated for visibility.

Performance is not inherently dishonest. Humans naturally adapt behavior to context. We speak differently in professional meetings than at home. We dress differently for formal events than for leisure. The difference lies in degree and persistence.

Online, performance becomes continuous. There is no clear boundary between private and public spheres. Content can be shared instantly with vast audiences. The awareness of potential visibility influences expression at every step.

The presence of metrics intensifies performance pressure. Likes, comments, shares, and views provide immediate feedback. These signals quantify reception. Quantification transforms self-expression into measurable output.

When output becomes measurable, comparison follows. Individuals monitor engagement relative to others. Success is interpreted through numbers. The self is evaluated publicly.

This evaluation can alter motivation. People may post content optimized for engagement rather than authenticity. Trends influence creative decisions. Algorithmic preferences shape tone, style, and subject matter.

The feedback loop between platform and user reinforces this adaptation. Content that performs well is rewarded with greater reach. Greater reach encourages repetition of similar themes. Over time, expression aligns with engagement patterns.

The performance of self extends beyond creative spaces. Everyday users also adapt behavior to perceived audience expectations. Opinions may be moderated to avoid backlash. Interests may be emphasized to align with trending culture.

This self-monitoring can become habitual. The internal question shifts from "Is this true for me?" to "Will this be received well?" External validation becomes a guiding principle. Authentic impulses may be filtered before expression.

Psychologically, performance requires cognitive resources. Managing impressions demands attention. Deciding how to present oneself consumes mental energy. Continuous self-curation contributes to fatigue.

In professional contexts, the pressure to maintain a polished online presence can blur boundaries. Résumés extend into social feeds. Personal and professional identities merge. The distinction between private individual and public brand diminishes.

For content creators and influencers, performance becomes livelihood. Income may depend on engagement rates. Authenticity competes with marketability. Personal experiences transform into monetizable narratives.

This commodification encourages selective sharing. Only certain aspects of life are displayed. Challenges may be framed carefully to maintain audience appeal. Vulnerability becomes strategic.

While curated sharing can inspire others, excessive performance can distance the self from lived reality. The gap between presentation and experience may widen. Maintaining consistency across contexts requires emotional labor.

Young users are especially influenced by performative norms. Adolescents observe models of identity construction online. They learn that visibility equates to value. They may equate popularity with self-worth.

The pressure to perform can discourage experimentation. Fear of judgment limits expression. Creative risks may feel unsafe. Conformity becomes the easier path.

Algorithms contribute to this environment by rewarding specific types of content. Visual aesthetics, dramatic narratives, and emotionally resonant posts often receive more engagement. Users adapt accordingly. Identity becomes shaped by algorithmic feedback.

The performance of self can also fragment identity. Different platforms encourage different tones. A professional network emphasizes competence. A visual platform emphasizes appearance. A discussion forum emphasizes opinion.

Navigating multiple personas requires flexibility. However, constant adaptation may create internal dissonance. Aligning external presentation with internal values demands awareness.

The psychological cost of performance increases when feedback fluctuates unpredictably. Posts that once performed well may later receive minimal attention. The inconsistency of engagement introduces uncertainty. Uncertainty can heighten stress.

The desire for recognition is natural. Humans seek acknowledgment from peers. Social validation strengthens bonds. Digital platforms magnify this desire through quantification.

Reducing performance pressure involves reframing motivation. Sharing should be guided by intention rather than reaction. Expression can prioritize meaning over metrics. Value can be defined internally.

Design changes could reduce the emphasis on public metrics. Hidden like counts or private feedback systems might decrease comparison. Emphasizing conversation over broadcasting can shift dynamics.

Offline relationships help ground identity in unperformed interactions. Face-to-face exchanges reveal unfiltered selfhood. In-person presence requires less strategic presentation. Authenticity emerges more naturally.

Practices such as journaling can support internal alignment. Writing privately allows exploration without audience consideration. Reflection clarifies values independent of response.

Digital boundaries also reduce performance strain. Limiting the frequency of posting lowers pressure to maintain constant presence. Scheduled breaks from platforms restore balance.

Communities that prioritize authenticity over virality can reshape norms. Encouraging depth, empathy, and honesty reduces performative escalation. Cultural expectations influence behavior significantly.

The performance of self reflects broader societal shifts toward visibility culture. In a world where attention is scarce and valuable, appearing becomes important. Yet presence without performance fosters stability.

Identity should not depend solely on external applause. Internal coherence provides resilience. When values guide behavior, engagement metrics lose dominance.

Technology can facilitate authentic expression when designed with care. Platforms that emphasize meaningful dialogue rather than popularity contests can support healthier identity formation. Transparency about algorithmic influence empowers users.

The neon glow of digital spaces can tempt individuals to craft idealized versions of themselves. But sustainable well-being requires integration between public and private selves. Consistency across contexts strengthens psychological health.

The performance economy thrives on constant output. Resisting it requires deliberate restraint. Choosing silence at times can be as meaningful as sharing. Absence can protect integrity.

The performance of self does not have to define modern identity. By recognizing its influence, individuals can decide when to engage per-formatively and when to withdraw. Conscious choice restores balance.

In the afterworld shaped by algorithms and metrics, authenticity becomes a form of quiet resistance. When expression aligns with values rather than visibility, identity stabilizes. And stability provides a foundation for deeper connection and growth.

CHAPTER . 14

Burnout in a 24/7 World

There was a time when work had clear boundaries. Offices closed at defined hours. Messages waited until the next day. Rest was structurally protected by the limits of physical space and analog communication. Productivity had a rhythm. So did recovery.

The digital world dissolved those boundaries. Emails arrive at all hours. Notifications interrupt evenings and weekends. Remote access enables constant availability. Work can follow individuals home through the devices they carry in their pockets.

Availability has quietly transformed into expectation. Being reachable is often interpreted as being committed. Rapid response can signal professionalism. Delayed replies may be judged as disengagement. The cultural norm shifts toward immediacy.

This expectation creates cognitive strain. The mind rarely fully disengages. Even during rest, it may anticipate incoming tasks. The boundary between labor and leisure becomes blurred. Without separation, recovery diminishes.

Burnout develops gradually under sustained pressure. It is characterized by emotional exhaustion, reduced performance, and diminished motivation. In a 24/7 environment, recovery windows shrink. Continuous connectivity prevents full restoration.

The human nervous system requires cycles of activation and calm. Stress responses are useful in short bursts. They mobilize energy for deadlines and challenges. But without sufficient downtime, stress becomes chronic.

Chronic stress affects sleep quality, immune function, and mood regulation. Persistent alertness keeps the body in heightened readiness. Relaxation becomes harder to access. Over time, fatigue accumulates.

Digital tools intensify this condition by enabling multitasking. Switching between applications fragments attention. Responding to messages while completing other tasks reduces efficiency. Cognitive switching consumes mental resources.

The pressure to remain productive can extend beyond formal work hours. Side projects, personal branding, and continuous learning all compete for time. Optimization culture promotes constant self-improvement. Rest may feel unproductive.

Yet rest is not idleness. It is a biological necessity. Downtime consolidates memory and restores mental clarity. Without rest, creativity declines. Decision-making deteriorates.

Burnout is not merely individual weakness. It reflects structural conditions. When systems reward constant output, individuals adapt accordingly. If performance metrics prioritize volume over sustainability, exhaustion follows.

Remote work environments highlight this dynamic. While flexibility offers benefits, it also removes physical transitions between roles. Commuting once provided psychological separation. Now workspaces often overlap with living spaces.

The absence of spatial boundaries requires intentional temporal boundaries. Without conscious scheduling, tasks can expand indefinitely. The workday risks extending continuously.

Notifications are central to this expansion. Real-time alerts demand attention. Even silent alerts can create anticipatory stress. The possibility of interruption affects concentration.

This anticipatory stress consumes mental bandwidth. The brain allocates resources toward monitoring potential messages. Focus diminishes. Deep engagement becomes more difficult.

Burnout also intersects with performance culture. Public metrics, deadlines, and visibility amplify pressure. When output is tracked and compared, competition intensifies. Self-worth may become linked to productivity.

In creative fields, algorithms can reinforce this pattern. Content must be produced consistently to maintain visibility. Irregular posting can reduce reach. Stability depends on frequency.

This cycle encourages overextension. Individuals push beyond sustainable limits to meet perceived demands. Short-term gains may come at long-term cost. Health deteriorates gradually.

The 24/7 world erodes recovery rituals. Sleep schedules shift. Meal times vary. Leisure blends with work-related browsing. The day lacks clear structure.

Without structure, the nervous system struggles to downshift. Predictable routines help regulate stress. Consistent boundaries signal safety. Irregular engagement prolongs activation.

Organizations can counter burnout by establishing communication norms. Clear expectations about response times reduce pressure. Encouraging offline periods improves sustainability. Leadership behavior influences culture.

Individual strategies are also essential. Setting device-free hours restores mental space. Turning off nonessential notifications reduces interruptions. Creating separate work environments reinforces boundaries.

Physical movement supports recovery. Exercise reduces stress hormones and improves mood. Outdoor time reconnects individuals with natural rhythms. Exposure to daylight stabilizes sleep cycles.

Community support mitigates burnout as well. Shared workloads distribute responsibility. Collaborative environments reduce isolation. Social connection buffers stress.

Digital minimalism practices emphasize intentional usage. By evaluating which platforms serve genuine goals, individuals can reduce unnecessary engagement. Eliminating low-value interactions frees time for rest.

Burnout in the digital era is often invisible. From the outside, constant online presence can appear productive. Internally, however, energy may be depleted. The difference between activity and sustainability is significant.

Recovery requires more than temporary breaks. It requires systemic recalibration. Work culture, technology design, and personal habits must align. Sustainability must be prioritized alongside efficiency.

The neon environment rewards immediacy. But human physiology rewards balance. When we honor biological limits, performance improves. When we ignore them, decline follows.

Burnout signals misalignment between demands and capacity. Recognizing this signal is the first step toward change. Ignoring it deepens exhaustion.

In a world that never sleeps, rest becomes an intentional act. Scheduling downtime protects cognitive health. Treating recovery as essential rather than optional shifts perspective.

Technology can support balance when configured thoughtfully. Calendar boundaries, focus modes, and notification controls can reduce intrusion. Platforms can design features that encourage breaks.

The 24/7 paradigm does not have to define the future. Societies can choose rhythms that respect human limits. Economic systems can reward sustainability instead of constant output.

Burnout is not inevitable. It is the result of cumulative pressures interacting with accessible tools. By restoring boundaries, we restore capacity.

The afterworld's glow can be constant, but human life requires cycles. Work and rest, engagement and withdrawal, action and recovery. When these cycles are respected, resilience grows.

Reclaiming rest is not retreat. It is strategy. It preserves creativity, health, and long-term effectiveness. In doing so, it transforms exhaustion into endurance.

CHAPTER . 15

The Attention Crisis

Attention is the foundation of experience. Whatever we consistently focus on becomes our perceived reality. It shapes memory, emotion, learning, and identity. Attention determines what enters consciousness and what remains invisible. In many ways, attention is the most valuable resource a human possesses.

In the modern digital environment, attention is under constant competition. Notifications, advertisements, messages, and updates all vie for priority. Each platform seeks to capture and retain focus. The result is not just distraction, but fragmentation.

Fragmentation weakens depth. When attention shifts repeatedly in short intervals, sustained thought becomes difficult. Complex problems require uninterrupted concentration. Creativity thrives in prolonged engagement. Without it, insight struggles to emerge.

The architecture of digital platforms is designed around engagement metrics. Time spent, clicks, and interactions are measured continuously. Algorithms learn which stimuli hold attention longest. They adapt accordingly. The system evolves to maximize focus capture.

This optimization has consequences. Content that triggers immediate reaction often outperforms material that requires reflection. Sensational headlines, emotionally charged imagery, and rapid novelty dominate feeds. Subtlety competes in an unequal arena.

Over time, users adapt to these patterns. The brain becomes accustomed to rapid input. Slow tasks may feel uncomfortable. Reading long passages can require greater effort than before. Patience diminishes as stimulation increases.

The attention crisis is not simply about distraction. It is about capacity. Deep attention enables learning, empathy, and strategic thinking. When attention is constantly divided, these abilities weaken. The quality of thinking declines.

Educational environments increasingly reflect this shift. Students raised in high-stimulation digital ecosystems may struggle with sustained focus. Traditional lecture formats can feel demanding. Short-form content reshapes expectations about pacing.

Work environments also feel the impact. Multitasking becomes normalized. Email interruptions break workflow. Messaging platforms encourage immediate response. Concentration fragments into intervals.

The cognitive cost of switching tasks is significant. Each transition requires mental reorientation. Even brief interruptions can reduce efficiency. Restoring deep focus takes time. Without structured boundaries, productivity suffers.

The attention economy treats focus as inventory. Platforms compete for it. Advertisers pay for access to it. Users supply it continuously. Yet unlike financial currency, attention is finite and renewable only through rest.

Chronic overextension of attention leads to fatigue. Mental exhaustion reduces clarity. Decision-making becomes reactive rather than reflective. Emotional regulation weakens under strain.

The crisis also affects relationships. Divided attention during conversations signals reduced presence. Checking devices mid-interaction diminishes connection. Over time, relational depth may erode.

Attention shapes moral judgment as well. What we focus on influences what we consider important. Repeated exposure to certain narratives can alter priorities. If outrage dominates attention, perception of the world may skew toward threat.

The algorithmic feedback loop reinforces this pattern. Engagement-driven systems amplify content that sustains attention. Users provide behavioral signals. The system responds. The cycle intensifies.

This dynamic can create a sense of urgency across all domains. News appears constantly updated. Social interactions feel immediate. Waiting becomes rare. The expectation of instant information permeates daily life.

But attention cannot operate at maximum intensity indefinitely. The brain requires periods of low stimulation to consolidate information. Downtime supports memory and insight. Without it, overload accumulates.

The attention crisis also intersects with well-being. Mindful presence correlates with reduced stress. When attention is scattered, anxiety may increase. Continuous input leaves little space for reflection.

Addressing the crisis begins with awareness. Recognizing when attention is being diverted allows conscious redirection. Identifying triggers helps reduce automatic responses. Intentional focus strengthens cognitive control.

Practical strategies can restore attention capacity. Setting dedicated time blocks for single tasks improves depth. Turning off nonessential notifications reduces interruption. Creating device-free zones protects concentration.

Engaging in activities that require sustained effort, such as reading, writing, or practicing a skill, rebuilds focus stamina. Attention functions like a muscle. It strengthens through deliberate use. It weakens through neglect.

Nature exposure also supports attentional recovery. Environments with lower stimulation allow the mind to rest. Natural settings reduce cognitive overload. Even short periods outdoors can improve clarity.

Organizations can contribute by reducing unnecessary meetings and digital interruptions. Clear communication guidelines help protect focus time. Cultural norms that respect concentration enhance collective performance.

Design innovation could support healthier attention patterns. Interfaces might include built-in pauses or prompts encouraging reflection. Algorithms could prioritize long-form content alongside short-form stimuli. Transparency about recommendation logic increases awareness.

The attention crisis reflects a mismatch between human cognitive limits and technological abundance. We have access to vast information, but limited processing capacity. Without boundaries, abundance becomes overwhelming.

The neon glow of devices is continuous, but human focus is cyclical. Alternating between engagement and rest is essential. Intentional rhythm restores balance.

Reclaiming attention is foundational to solving broader challenges discussed throughout this book. Without focused awareness, reform efforts lack clarity. Without sustained concentration, systemic change becomes difficult.

Attention determines the direction of collective energy. When it is dispersed, progress fragments. When it is aligned, transformation accelerates. Protecting attention is therefore both personal and societal responsibility.

The attention crisis is not irreversible. It is the result of design choices and cultural habits. By reshaping incentives and cultivating discipline, focus can be restored.

In the afterworld shaped by algorithms, attention is the key to autonomy. Whoever controls focus influences perception. By reclaiming control over attention, individuals regain agency over experience itself.

The solution begins with deliberate practice. It continues through design reform. And it culminates in a cultural shift that values depth over distraction. Only then can we move beyond crisis toward clarity.

CHAPTER . 16

The Infantilization of Society

Societies mature when individuals are encouraged to think independently, delay gratification, and take responsibility for consequences. Maturity requires patience, discipline, and tolerance for complexity. These qualities develop gradually through experience, education, and social expectation. They are reinforced when systems reward thoughtful engagement over impulsive reaction.

The digital environment, however, often rewards immediacy. Instant feedback loops encourage quick responses. Content designed for rapid consumption reduces the need for sustained focus. Short-form formats compress ideas into fragments. Over time, this environment can shape cognitive habits.

Infantilization in this context does not imply loss of intelligence. It refers to shifts in attention span, emotional regulation, and decision-making patterns. When systems prioritize novelty, simplicity, and constant stimulation, users may adapt accordingly. The brain adjusts to its environment.

Short-form content platforms exemplify this dynamic. Brief videos, rapid transitions, and endless feeds provide continuous micro-stimulation. The pace leaves little room for reflection. Complexity competes with entertainment. Depth must justify itself within seconds.

This structure can influence expectations beyond entertainment. Users may begin anticipating immediate clarity in discussions. Nuanced arguments can feel tedious. Extended explanations may struggle to retain attention. The preference for brevity extends into education, work, and civic discourse.

Emotional expression also shifts in highly reactive environments. Quick replies encourage spontaneous reactions. Thoughtful pauses may be interpreted as disengagement. The norm becomes responsiveness rather than deliberation. Impulse gains prominence.

In some cases, platforms simplify interfaces to maximize accessibility. While simplicity can improve usability, excessive simplification may reduce engagement with complexity. Decision-making becomes guided by pre-selected options. Autonomy can feel constrained by design.

Gamification elements contribute to infantilizing tendencies. Points, streaks, badges, and reward animations simulate achievement systems. These mechanisms can motivate positive behavior. However, when overused, they may shift motivation from intrinsic to extrinsic.

The constant presence of rewards and notifications mirrors conditioning structures used in early learning environments. While reinforcement can support habit formation, excessive external rewards may reduce internal regulation. Dependence on external validation can increase.

The attention economy amplifies this pattern. Platforms compete by offering increasingly engaging stimuli. If one format becomes popular, others adapt to match its brevity and intensity. The overall cultural tempo accelerates.

Acceleration can reduce tolerance for delay. Waiting becomes uncomfortable. Patience feels inefficient. Yet many meaningful outcomes require sustained effort over time. Skills, relationships, and institutions develop gradually.

Infantilization also appears in information consumption. Headlines replace in-depth articles. Summaries replace full reports. Commentary replaces analysis. This compression simplifies understanding but may obscure nuance.

When information is consistently delivered in simplified forms, cognitive habits adapt. Critical evaluation requires practice. If users rarely engage with complex material, analytical skills may weaken. Intellectual endurance diminishes.

The phenomenon is reinforced by algorithmic curation. Content that performs well in short engagement windows is prioritized. Detailed discussions may receive less visibility. The system adapts to consumption patterns, not necessarily educational value.

This dynamic affects public discourse. Complex policy debates are reduced to slogans. Nuanced positions struggle for attention. Emotional framing dominates rational explanation. The result can resemble simplified narratives rather than layered understanding.

Infantilization does not mean individuals lose capability. It suggests that environmental design influences behavior. When systems make depth inconvenient, many will choose convenience. When stimulation is abundant, restraint becomes less common.

Sleep disruption contributes as well. Late-night device use shortens restorative cycles. Fatigue reduces impulse control. Reduced rest can impair judgment. Over time, cognitive resilience declines.

Children growing up in highly stimulated environments may experience developmental impacts. Early exposure to rapid transitions can shape attention preferences. Educational structures must adapt to balance stimulation with focus-building exercises.

Parents and educators play crucial roles in countering infantilizing pressures. Structured routines encourage responsibility. Clear boundaries around device use foster self-regulation. Encouraging problem-solving rather than immediate answers strengthens independence.

Workplaces can also resist infantilization by promoting autonomy and critical thinking. Allowing employees to engage deeply with tasks without constant interruption enhances capability. Trust-based cultures reinforce maturity.

Design improvements could reintroduce friction where appropriate. Optional confirmation steps, customizable notification settings, and default limits on continuous playback may help users maintain control. Transparency about engagement metrics can increase awareness of usage patterns.

Cultural values influence the trajectory as well. Societies that celebrate discipline, curiosity, and perseverance will reinforce maturity. Societies that prioritize constant entertainment may drift toward superficial engagement.

Infantilization can also emerge when users rely heavily on algorithmic recommendations. Outsourcing decision-making reduces practice in independent evaluation. When systems suggest content, purchases, or opinions, autonomy may weaken subtly.

Reclaiming maturity involves reintroducing challenge. Reading long-form texts, engaging in sustained debate, and pursuing complex skills rebuild cognitive endurance. Effort strengthens mental resilience.

Community environments can reinforce adult responsibility by encouraging participation rather than passive consumption. Collaborative projects, volunteer efforts, and civic engagement foster ownership. Shared responsibility counters passivity.

The neon world promises effortless satisfaction. Immediate entertainment. Instant validation. Rapid information. But human growth requires friction. Without challenge, development stagnates.

Infantilization is not an accusation; it is a warning about structural incentives. Systems optimized purely for engagement may inadvertently discourage depth. Recognizing this tendency enables correction.

The attention crisis and burnout discussed earlier connect directly to this chapter's theme. When stimulation is constant and rest is limited, cognitive maturity suffers. Balance restores strength.

Society thrives when individuals are trusted with complexity. Encouraging thoughtful dialogue, extended reading, and long-term planning supports collective progress. Simplicity has its place, but not at the expense of substance.

Technology can either flatten or elevate experience. It can encourage dependence or independence. The difference lies in design philosophy and user intention.

If we wish to avoid societal infantilization, we must cultivate environments that reward patience, critical thinking, and sustained engagement. Depth must become valuable again.

The afterworld can either reduce citizens to passive consumers of content or support them as capable participants in shaping culture. The choice depends on incentives, education, and personal discipline.

Maturity in the digital age is not about rejecting technology. It is about using it with awareness and restraint. By reintroducing boundaries, embracing complexity, and prioritizing long-form engagement, we strengthen individual and societal capacity.

The glow of screens does not have to dim our growth. With intentional design and conscious use, technology can support rather than undermine development. Restoring balance ensures that innovation enhances adulthood instead of replacing it with perpetual stimulation.

CHAPTER . 17

Surveillance as Business Model

Surveillance once evoked images of governments monitoring citizens in secret. It implied hidden cameras, intelligence agencies, and state power. In the digital age, surveillance has shifted from a political tool to an economic foundation. It is no longer an exception. It is infrastructure.

The modern internet runs on data collection. Every click, pause, search, and interaction generates information. This information is recorded, analyzed, and transformed into behavioral insights. Those insights predict future actions. Prediction is the product.

The business model of many major platforms depends on attention and behavioral data. Users receive free services in exchange for participation. Participation means continuous engagement and data generation. The more time users spend online, the more signals they produce.

These signals reveal patterns. Location data shows movement habits. Search queries expose curiosity and intent. Interaction histories reveal preferences and relationships. Even brief pauses can indicate interest. Taken together, these signals form detailed profiles.

Surveillance in this context is not necessarily malicious in intent. It is structural. The architecture of advertising-driven platforms rewards data extraction. Revenue increases with predictive accuracy. Predictive accuracy improves with more data.

This feedback loop creates incentives for expansion. More tracking tools are introduced to refine personalization. Cookies, pixels, device identifiers, and cross-platform integrations enhance visibility. Each innovation increases resolution.

Users often consent to data collection through lengthy agreements. Terms and conditions outline practices in technical language. Most individuals accept without reading. Convenience outweighs scrutiny. The transaction feels abstract rather than immediate.

Over time, data accumulation becomes normalized. Continuous monitoring is reframed as customization. Personalized recommendations are presented as benefits. Surveillance is softened through language.

The scale of data aggregation is unprecedented. Information from multiple sources can be combined. Data brokers compile records from public and private channels. Profiles become increasingly comprehensive.

These profiles inform targeted advertising. Advertisers can reach specific demographics with precision. Campaigns can be tailored to interests, behaviors, and predicted vulnerabilities. Efficiency improves.

Surveillance-based models extend beyond advertising. Credit scoring systems analyze financial behavior. Health applications track biometric data. Smart devices monitor household activity. Each domain introduces new forms of data exchange.

The aggregation of data also raises security risks. Breaches expose sensitive information. Large datasets become attractive targets for cyberattacks. Vulnerabilities increase with scale.

Trust becomes central to sustainability. If users lose confidence in data stewardship, engagement may decline. Transparency about collection practices can mitigate skepticism. Clear communication fosters accountability.

However, the economic incentives for data retention remain strong. Data represents competitive advantage. Companies with larger datasets can refine algorithms more effectively. This advantage can consolidate market power.

Concentration of data can lead to concentration of influence. Organizations with access to extensive behavioral information can shape advertising, content visibility, and product design. Their impact extends across sectors.

The normalization of surveillance also influences cultural expectations. Individuals may become accustomed to being tracked. Location sharing, wearable devices, and public posting create continuous digital footprints. Privacy boundaries evolve.

Privacy does not mean secrecy. It means control over personal information. When individuals lack meaningful choice about data usage, autonomy diminishes. Control becomes asymmetrical.

Regulatory frameworks attempt to address these concerns. Data protection laws require disclosure and consent. Some regions mandate transparency and user rights. Enforcement, however, varies.

Ethical design principles can reduce intrusive practices. Minimizing data collection to what is necessary limits exposure. Storing information securely protects users. Designing with privacy by default supports trust.

Surveillance as a business model shapes platform behavior. If revenue depends on data extraction, incentives favor engagement maximization. Features that increase time spent online align with profit goals. Design decisions reflect this alignment.

This alignment influences content ecosystems. Algorithms optimize for retention. Retention benefits from personalized feeds.

Personalization requires data analysis. Surveillance underpins the entire structure.

Users may experience this indirectly through highly tailored recommendations. While personalization can feel convenient, it also reveals how extensively behavior is monitored. The precision of suggestions implies depth of tracking.

The psychological effect can vary. Some users appreciate relevance. Others feel discomfort at perceived intrusiveness. Awareness of tracking may alter behavior. Self-censorship can occur.

In democratic societies, surveillance raises additional questions. Information environments shape public opinion. When platforms control data-driven visibility, they influence discourse. Transparency in ranking systems becomes critical.

Educational efforts in digital literacy can empower users to understand data flows. Learning how information is collected and monetized demystifies processes. Informed users can make deliberate choices about participation.

Surveillance models are not inevitable. Alternative business structures exist. Subscription-based platforms rely less on data extraction. Cooperative ownership models can prioritize user interests. Policy incentives can encourage diversification.

Technological innovation can support privacy-enhancing methods. Techniques such as data minimization and decentralized storage reduce centralized accumulation. Encryption strengthens security. Responsible engineering can reshape outcomes.

The neon infrastructure of the digital economy often hides its surveillance foundation. Interfaces emphasize convenience and creativity. The underlying systems depend on continuous monitoring. Recognizing this structure clarifies power dynamics.

Surveillance as a business model represents a fundamental shift in economic organization. Information about behavior has become more valuable than the behavior itself. Predicting action generates profit. Data fuels prediction.

The challenge is balancing innovation with rights. Data-driven services offer benefits in navigation, healthcare, and personalization. Eliminating data use entirely is neither realistic nor necessarily beneficial. The goal is proportionality.

Proportionality requires consent that is informed and revocable. It requires transparency about usage. It requires safeguards against misuse. It requires accountability.

When surveillance becomes excessive, it can erode trust and autonomy. When it is constrained by ethical frameworks, it can coexist with innovation. Governance shapes direction.

In the afterworld shaped by data flows, understanding surveillance is essential. Without awareness, individuals cannot assess trade-offs. With awareness, they can participate consciously.

Surveillance as a business model is powerful because it operates quietly. It does not rely on force. It relies on participation. Users generate data through everyday activity.

Recognizing this reality enables informed engagement. It encourages questioning default settings. It promotes consideration of privacy tools. It inspires demand for transparent systems.

The digital gods described earlier thrive on information. Surveillance provides their sight. By understanding this mechanism, we gain insight into how influence is structured.

If societies choose to prioritize privacy and ethical design, surveillance can be moderated. If incentives remain unchanged, expansion will likely continue. The future depends on collective decisions.

The machine is built. The question is how it will be governed. Awareness is the first step toward responsible stewardship of data in the modern economy.

Chapter . 18

Democracy in the Algorithmic Era

Democracy depends on informed participation. Citizens must have access to reliable information in order to evaluate candidates, policies, and institutions. Public debate must occur within a shared framework of facts. When that framework weakens, democratic processes become strained.

In the algorithmic era, information flows are mediated by platforms that rank, filter, and personalize content. These systems determine what users see first, what gains prominence, and what fades from visibility. The structure of these rankings influences public discourse at scale.

Algorithms are designed to optimize engagement. Engagement often correlates with emotional intensity. Content that provokes strong reactions tends to spread more rapidly. As a result, the informational environment may favor sensational material over measured analysis.

This dynamic can shape political communication. Campaigns adapt messaging to perform well within algorithmic systems. Short statements, striking visuals, and emotionally resonant themes receive priority. Nuanced policy discussions may struggle to compete.

Voters increasingly encounter political content through social feeds rather than traditional journalism. The feed integrates news, commentary, advertisements, and entertainment in a single stream. Context can become fragmented. Distinctions between reporting and opinion may blur.

Personalization intensifies this effect. Different users receive different versions of the political landscape. Search results and recommendations vary based on past behavior. Two individuals may perceive divergent realities when examining the same topic.

This divergence complicates consensus-building. Democratic systems rely on some degree of shared understanding. When informational environments differ significantly, common ground narrows. Dialogue becomes more difficult.

Algorithmic amplification can also accelerate polarization. Content that triggers disagreement generates comments and shares. Engagement metrics reward divisive framing. Over time, the system may surface increasingly extreme perspectives.

Political actors may exploit these incentives. Messaging strategies can be tailored to specific demographic groups using data analytics. Micro-targeted advertisements deliver customized appeals. This precision enhances persuasion efficiency.

While targeted communication is not inherently problematic, opacity in targeting can limit public oversight. Citizens may not know which messages others receive. Transparency about political advertising practices becomes important for accountability.

The speed of digital dissemination further transforms democratic processes. Information spreads rapidly across networks. Corrections may not travel as widely as initial claims. False or misleading narratives can gain traction before verification occurs.

This rapid spread can influence public opinion during critical moments. Elections, referendums, and policy debates unfold within compressed timeframes online. Momentum generated by viral content can affect perceptions.

Algorithmic systems also influence visibility of civic information. Official announcements, public service updates, and educational materials compete with entertainment content. Ranking priorities determine reach. Design choices impact exposure.

Trust in institutions is closely tied to informational reliability. When users encounter conflicting narratives across platforms, confidence may decline. Perceived inconsistency can undermine belief in governance structures.

However, algorithms are not inherently anti-democratic. They can also improve access to diverse viewpoints if configured intentionally. Recommendation systems can be designed to broaden exposure rather than narrow it. Transparency in design increases public trust.

Democracy in the algorithmic era requires adaptation. Traditional media literacy must expand to include understanding of platform mechanics. Citizens benefit from knowledge about how content is selected and amplified.

Educational systems can incorporate instruction on algorithmic influence. Learning how ranking systems function enables critical engagement. Awareness reduces susceptibility to manipulation.

Regulatory frameworks may also evolve to address platform responsibility. Requirements for transparency in political advertising, data usage, and content moderation can strengthen accountability. Balancing free expression with harm reduction remains complex.

Digital public squares now host much of civic discourse. Their governance therefore has societal implications. Decisions about moderation policies affect speech dynamics. Clear guidelines and consistent enforcement support stability.

Community norms play a role as well. Encouraging respectful dialogue and evidence-based argumentation fosters healthier debate. Public culture shapes online interaction patterns.

The algorithmic era introduces new challenges for election integrity. Disinformation campaigns can leverage automated accounts to amplify narratives. Coordinated networks may simulate grassroots movements. Detecting such manipulation requires technical vigilance.

Despite these challenges, digital tools also enable civic engagement. They facilitate mobilization, fundraising, and awareness campaigns. Grassroots movements can organize rapidly. Information can reach wide audiences efficiently.

The impact of algorithms on democracy depends largely on incentives and oversight. Systems optimized solely for engagement may inadvertently intensify division. Systems designed with civic values in mind can support informed participation.

User agency remains significant. Individuals can diversify their information sources intentionally. Following multiple reputable outlets reduces echo chamber effects. Cross-referencing claims enhances reliability.

Platform design choices matter greatly. Features that encourage contextual information, source labeling, and diverse recommendations can strengthen discourse. Reducing algorithmic opacity improves accountability.

Democracy requires not only access to information but also deliberative spaces. Structured forums for discussion, both online and offline, enable thoughtful exchange. Moderated environments can support productive dialogue.

The speed of the algorithmic environment often contrasts with the pace of democratic processes. Laws and policies develop gradually. Technological change occurs rapidly. Bridging this gap is essential.

The neon infrastructure of digital systems shapes public consciousness. Attention flows influence perception of importance. When algorithms prioritize certain topics, those topics dominate discourse. Visibility influences agenda-setting.

Safeguarding democracy in this context involves protecting informational integrity. This includes combating misinformation, ensuring transparency, and promoting media literacy. It also includes aligning platform incentives with civic health.

Democracy has always evolved alongside communication technologies. The printing press, radio, and television each transformed public discourse. Algorithms represent the latest shift. Adaptation is necessary.

The algorithmic era presents both risk and opportunity. It can fragment reality or connect communities. It can amplify division or broaden understanding. Outcomes depend on governance, design, and participation.

Citizens, institutions, and technology providers share responsibility. Collaboration is essential to maintain informed deliberation. Continuous evaluation ensures systems remain aligned with democratic values.

By understanding how algorithms influence visibility, we gain insight into their political impact. Awareness empowers voters to interpret information critically. Informed engagement strengthens resilience.

Democracy in the algorithmic era is not predetermined. It is shaped by choices about transparency, regulation, and user behavior. Conscious design and conscious consumption together can preserve democratic integrity within digital systems.

CHAPTER . 19

Deepfakes and Synthetic Reality

Truth has always depended on evidence. In earlier eras, photographs, recordings, and eyewitness accounts served as anchors of credibility. While imperfect, these forms of documentation carried persuasive weight. Seeing and hearing were closely tied to believing.

Advances in artificial intelligence have disrupted this relationship. Modern systems can generate highly realistic images, audio, and video that depict events that never occurred. These synthetic creations are increasingly difficult to distinguish from authentic media. The boundary between documentation and fabrication is weakening.

Deepfake technology uses machine learning models trained on large datasets of human faces, voices, and movements. By analyzing patterns, these systems can simulate expressions and speech with remarkable accuracy. The resulting media can appear convincingly real to casual viewers.

This capability introduces both creative and ethical implications. In entertainment and filmmaking, synthetic media can reduce production costs and enable new artistic techniques. Historical figures can be recreated digitally for educational purposes. Language translation can be enhanced with voice synthesis.

However, the same technology can be used maliciously. Fabricated videos can depict individuals saying or doing things they never did. False evidence can spread rapidly online. Reputations can be damaged before verification occurs.

The speed of digital dissemination amplifies the risk. A convincing synthetic video can circulate globally within hours. Emotional reactions often precede fact-checking. Corrections may not reach the same audience as the original content.

Synthetic audio presents additional challenges. Voice-cloning systems can replicate speech patterns with minimal sample data. This enables impersonation in contexts such as financial fraud or political manipulation. Trust in recorded communication becomes more complex.

As realism improves, skepticism may increase broadly. When users cannot easily differentiate authentic from synthetic media, uncertainty grows. This phenomenon is sometimes described as the erosion of epistemic trust. If people doubt all media equally, distinguishing truth from falsehood becomes harder.

This environment creates opportunities for misinformation campaigns. Actors can produce fabricated content to support false narratives. Deepfakes can be strategically released to influence public opinion. The perceived authenticity of such media enhances persuasive power.

The potential for misuse has prompted research into detection methods. Technical tools analyze inconsistencies in lighting, pixel patterns, and audio artifacts. However, as generation models improve, detection methods must continuously evolve. The interaction resembles an ongoing technological arms race.

Legal systems are beginning to address synthetic media. Some jurisdictions consider regulations targeting malicious deepfake distribution. Policy discussions focus on balancing free expression with protection against harm. Enforcement remains complex in a global digital environment.

Public awareness plays a crucial role in resilience. Education about synthetic media helps individuals approach unexpected content critically. Understanding that realistic fabrication is possible reduces automatic belief. Media literacy becomes increasingly important.

Platforms can implement labeling mechanisms for AI-generated content. Clear disclosure can help users distinguish between authentic and synthetic material. Transparency reduces ambiguity. Consistent standards strengthen trust.

Watermarking techniques are also being explored. Embedded digital signatures may indicate origin and authenticity. Such systems require widespread adoption to be effective. Cooperation among technology providers enhances impact.

Despite the risks, synthetic media also holds constructive potential. Educational simulations can recreate historical environments. Language learning tools can generate interactive scenarios. Accessibility technologies can assist individuals with disabilities.

The challenge lies in governance and intent. Tools themselves are neutral in capability. Their impact depends on how they are deployed. Ethical frameworks guide responsible usage.

Deepfakes also affect personal relationships. Individuals may fear that images or recordings could be manipulated. Trust in private communication may require additional safeguards. Verification processes become more common.

The psychological impact of synthetic reality extends beyond specific incidents. When users are aware that media can be fabricated convincingly, overall skepticism may increase. This skepticism can be healthy when applied carefully. However, excessive doubt can hinder collective agreement on facts.

Societies rely on shared evidence to function effectively. If public trust in documentation declines significantly, cooperation becomes more difficult. Institutions must adapt to maintain credibility in an environment of advanced synthesis.

Verification mechanisms may become standard practice. Multi-source confirmation, cryptographic authentication, and platform accountability can reduce confusion. Technical infrastructure must evolve alongside generative capabilities.

Deepfakes highlight the broader theme of this book: systems shape perception. Just as algorithms influence visibility, synthetic media influences belief. Both operate within the same digital ecosystem. Both affect reality as experienced by users.

The neon glow of modern devices can now display images that never existed. What once required physical staging can now be generated digitally. The cost of fabrication has decreased dramatically. As accessibility increases, vigilance becomes essential.

Synthetic reality challenges the assumption that seeing is believing. It encourages critical evaluation of sources. It underscores the importance of corroboration. Trust must be earned through transparency.

The future will likely include increasingly sophisticated generative models. This trajectory makes preparation vital. Technical defenses, legal frameworks, and public education must develop in parallel.

The presence of deepfakes does not mean truth is unattainable. It means verification must become more deliberate. Institutions, media organizations, and individuals share responsibility for maintaining informational integrity.

In the algorithmic era, authenticity requires infrastructure. Clear labeling, robust detection, and consistent standards form part of that infrastructure. Cultural norms also matter. Encouraging skepticism without cynicism balances awareness.

Deepfakes represent a turning point in the relationship between representation and reality. As synthetic media improves, society must refine its methods for confirming truth. Adaptation is necessary to preserve trust.

The challenge is not technology itself, but its alignment with ethical principles. With thoughtful governance and informed users, synthetic tools can serve creativity and innovation. Without safeguards, they can undermine credibility.

Understanding deepfakes is essential to navigating the afterworld of digital systems. Awareness transforms uncertainty into preparedness. By anticipating synthetic influence, societies can strengthen resilience and maintain confidence in evidence.

CHAPTER . 20

The Automation of Work

Work has always defined more than income. It shapes identity, structure, and purpose. For many people, occupation provides rhythm to daily life and meaning within community. Employment connects individuals to shared goals and collective progress. It anchors time.

The rise of automation is transforming this relationship. Machines once replaced primarily physical labor. Industrial equipment increased efficiency in manufacturing and logistics. Today, software systems and artificial intelligence extend automation into cognitive domains. Tasks involving analysis, communication, and prediction are increasingly assisted or performed by algorithms.

Automation is not inherently negative. It can increase safety, reduce repetitive strain, and improve productivity. In hazardous environments, machines can perform tasks too dangerous for humans. In administrative contexts, automation can streamline workflows and reduce errors. Efficiency gains can benefit societies broadly.

However, the scale and speed of current technological change introduce new challenges. As systems become capable of performing complex tasks, certain roles evolve or disappear. Routine cognitive work, data processing, and content generation can be partially automated. This shift alters labor markets.

Job displacement can create uncertainty for workers whose skills align with tasks being automated. Transition periods may require retraining or adaptation. The ability to reskill becomes increasingly important. Educational systems must respond to evolving demand.

Automation also reshapes expectations of productivity. When software can complete tasks rapidly, human output may be measured against machine speed. This comparison can generate pressure. The standard for efficiency rises continuously.

The integration of AI tools into workplaces changes collaboration patterns. Humans increasingly supervise or refine machine outputs. Roles shift toward oversight, interpretation, and strategic judgment. Rather than replacing humans entirely, many systems augment human capability.

Nevertheless, augmentation can still alter identity. When core tasks change significantly, workers may feel disconnected from prior expertise. Professional confidence may require adjustment. Psychological adaptation accompanies technical change.

Economic inequality may also be affected. Organizations that deploy advanced automation can reduce operational costs. This advantage may concentrate resources among technology-intensive firms. Market dynamics can favor entities with access to large-scale computational infrastructure.

Labor markets must therefore evolve in response. Investment in education and continuous learning becomes essential. Skills emphasizing creativity, critical thinking, empathy, and complex problem-solving remain valuable. These capabilities complement automated systems.

Automation extends beyond traditional office environments. Autonomous vehicles, robotic systems, and predictive algorithms influence transportation, healthcare, and finance. Decision-support tools assist in diagnostics and planning. The boundaries between human and machine input blur.

In healthcare, automation can enhance accuracy in image analysis and data interpretation. In logistics, routing algorithms optimize delivery efficiency. In finance, predictive models inform risk assessment. These applications can improve outcomes when properly regulated.

However, reliance on automated systems introduces dependency. When decisions are delegated to algorithms, oversight becomes crucial. Transparency in how systems generate outputs supports accountability. Understanding limitations prevents overconfidence.

The psychological dimension of automation deserves attention. If machines perform increasing portions of work, individuals may question the meaning of their contributions. Purpose derived from employment may shift. Societies may need to redefine value beyond traditional labor models.

Some proposals suggest universal basic income or alternative economic frameworks to address automation-driven changes. These discussions aim to ensure stability during transitions. Policy experimentation may become necessary as technology evolves.

Workplaces that integrate automation successfully often emphasize human strengths. Creativity, interpersonal communication, and strategic thinking are difficult to replicate fully. Designing roles that leverage these qualities enhances collaboration between humans and machines.

Continuous learning becomes central in an automated economy. Training programs, online courses, and professional development initiatives support adaptation. Lifelong education replaces static skill acquisition. Flexibility becomes a core competency.

The automation of work also intersects with surveillance-based business models. Data collected from work platforms can inform performance analytics. Monitoring tools track productivity metrics. While these systems can provide insights, they may also increase stress if used excessively.

Balance is key. Automation should reduce unnecessary burdens, not intensify monitoring without benefit. Clear policies on data usage within workplaces support trust. Employees need transparency about how their information is handled.

The future of work will likely involve collaboration with increasingly sophisticated AI systems. Rather than viewing automation as replacement, societies can frame it as partnership. Humans can focus on tasks requiring judgment, empathy, and innovation. Machines can handle repetitive and computational functions.

This transition will not be uniform across regions or industries. Some sectors will adapt quickly. Others may experience slower integration. Support systems must accommodate varied timelines.

Public dialogue about automation should emphasize preparation rather than fear. Understanding technological trends allows proactive planning. Governments, educational institutions, and private organizations can coordinate to anticipate shifts.

The automation of work represents one of the most significant structural changes in the digital age. It affects economic stability, identity, and social organization. Addressing its implications requires collaboration across disciplines.

In the broader context of this book, automation reflects the recurring theme of systems shaping human experience. Just as algorithms influence attention and synthetic media influences perception, automation influences labor and purpose. The interaction between design and behavior remains central.

Technology's role in work will continue to expand. The challenge lies in ensuring that expansion enhances human well-being. Ethical frameworks, education reform, and adaptive policy can guide transition.

If managed thoughtfully, automation can reduce drudgery and increase creative opportunity. If unmanaged, it may amplify inequality and uncertainty. The outcome depends on intentional governance.

Understanding automation in this way prepares societies for future developments. Awareness enables adaptation. Adaptation fosters resilience. Resilience ensures that technological progress aligns with human flourishing.

Chapter . 21

The Wealth Concentration Spiral

Economic systems have always produced winners and losers. Innovation creates opportunity, but it also reshapes competitive landscapes. In the digital era, however, the scale of reward has changed dramatically. Technology platforms can grow globally with minimal physical infrastructure. This scalability enables unprecedented expansion of wealth.

Digital businesses often rely on network effects. The more users a platform attracts, the more valuable it becomes. Increased value draws even more users. This self-reinforcing loop accelerates growth. Early advantages compound rapidly.

Unlike traditional industries that require factories, land, or extensive distribution chains, digital platforms can expand across borders almost instantly. Software replicates at negligible marginal cost. Once developed, it can serve millions without proportional increases in production expense. Efficiency at scale magnifies returns.

This structure contributes to wealth concentration. A small number of companies can dominate entire sectors. Market share consolidates around platforms that control data, infrastructure, and user attention. As dominance increases, competitive barriers rise.

Access to large datasets provides additional advantage. Machine learning systems improve with volume and diversity of information. Companies with extensive data repositories refine algorithms more effectively. Improved performance attracts more users, reinforcing dominance.

Capital investment further accelerates the spiral. Venture funding often prioritizes rapid growth over short-term profitability. Investors seek scalable returns. Platforms that capture large user bases become highly valuable assets. Acquisition activity can consolidate emerging competitors.

This cycle may lead to reduced market diversity. Smaller firms face challenges competing with established platforms. Innovation can still occur, but distribution channels may depend on dominant ecosystems. Integration into existing networks becomes necessary for visibility.

Wealth concentration has social implications. Economic inequality may widen when returns on digital capital outpace wage growth. Owners of platform infrastructure accumulate substantial resources. Workers and users contribute value indirectly through participation and data generation.

The attention economy intersects with this spiral. Users provide time and engagement, which translate into revenue streams. Engagement fuels advertising profits. Advertising revenue funds further expansion. The loop strengthens financial accumulation.

Global reach intensifies the effect. Digital platforms can operate across multiple countries simultaneously. Revenue streams are not confined by geography. Regulatory environments vary, but digital services transcend borders. This mobility enhances strategic advantage.

Wealth concentration also influences political dynamics. Corporations with substantial economic power may have resources to engage in lobbying and policy advocacy. Their interests intersect with regulatory decisions. Governance structures must balance innovation incentives with public accountability.

However, concentration does not eliminate opportunity. Digital ecosystems also enable entrepreneurship. Individuals can create content, applications, and services with relatively low initial capital. Platforms provide distribution channels that were previously inaccessible.

The key tension lies in dependency. When distribution relies heavily on a small number of platforms, access to markets may depend on algorithmic visibility. Changes in ranking systems can affect revenue. This dependency introduces vulnerability.

Diversification of infrastructure can mitigate concentration risks. Supporting open standards and interoperable systems encourages competition. Encouraging data portability allows users to move between services more easily. Competitive markets benefit from flexibility.

Wealth concentration also interacts with automation. Companies that invest heavily in AI and robotics can increase productivity significantly. Higher productivity can translate into greater profits. These profits may further fuel technological investment.

This compounding effect reinforces the spiral. Capital enables innovation. Innovation increases efficiency. Efficiency increases profit. Profit attracts more capital. The cycle accelerates.

At the societal level, concentration may influence access to resources such as education, healthcare, and housing. Economic disparities can affect opportunity distribution. Addressing inequality requires multifaceted policy approaches.

Tax systems, regulatory frameworks, and social programs play roles in balancing outcomes. The goal is not to suppress innovation, but to ensure shared benefit. Sustainable economies distribute value broadly.

Digital wealth generation differs from traditional manufacturing in that intangible assets dominate. Intellectual property, algorithms, and user networks form core value components. These assets are scalable and transferable. Their replication does not require physical expansion.

This characteristic enables rapid global dominance. Once a platform achieves critical mass, competitors face steep entry barriers. Users tend to remain within established ecosystems due to convenience and network presence. Switching costs reinforce retention.

The wealth concentration spiral is not inevitable, but it is structurally incentivized within current models. Network effects, data accumulation, and capital flow interact synergistically. Without counterbalancing mechanisms, concentration intensifies.

Policy discussions increasingly explore regulatory interventions. Antitrust measures, data governance standards, and transparency requirements aim to promote competition. These tools attempt to prevent excessive consolidation while preserving innovation.

Economic education can also empower individuals. Understanding how digital markets function allows more informed participation. Awareness of network dynamics clarifies why certain platforms dominate. Knowledge reduces opacity.

Community-based and cooperative digital initiatives offer alternative models. Platforms owned by users or stakeholders may distribute value differently. Decentralized systems can reduce concentration if widely adopted.

The spiral also intersects with labor shifts discussed earlier. As automation increases productivity, profits may rise even as workforce size remains stable or decreases. Aligning productivity gains with equitable distribution becomes essential for stability.

Wealth concentration influences societal resilience. Diversified economic systems are often more adaptable. Broad participation in value creation supports long-term sustainability. Inclusion strengthens cohesion.

The neon architecture of digital systems can amplify both innovation and inequality. The same scalability that enables creativity also enables consolidation. The direction depends on governance and market structure.

Recognizing the spiral provides an opportunity for proactive design. Encouraging competitive environments, supporting emerging innovators, and ensuring transparent practices can moderate concentration. Balanced ecosystems foster healthy growth.

The wealth concentration spiral is not merely an economic phenomenon. It reflects deeper structural patterns of digital scalability and data-driven advantage. Addressing it requires interdisciplinary collaboration between policymakers, technologists, and communities.

In understanding this dynamic, societies gain clarity about how digital wealth is generated and distributed. Awareness enables informed decisions about regulation, investment, and participation. With intentional action, technological progress can contribute to shared prosperity rather than excessive centralization.

CHAPTER . 22

Children of the Screen

Childhood has always been shaped by its environment. Playgrounds, neighborhoods, schools, and family structures define early development. Through exploration and interaction, children learn language, empathy, boundaries, and problem-solving. These foundations form the basis of adulthood.

Today, screens are among the most influential environments in childhood. Tablets, smartphones, televisions, and computers accompany children from increasingly early ages. Digital devices are present in homes, classrooms, and public spaces. Exposure begins earlier and lasts longer than in previous generations.

Screens offer educational opportunities. Interactive learning tools can reinforce literacy, numeracy, and creativity. Access to global information expands curiosity. Carefully designed content can support cognitive development. Technology itself is not inherently harmful.

However, the intensity and duration of exposure matter. Early childhood development relies heavily on sensory interaction and physical engagement. Touch, movement, and face-to-face communication stimulate neural growth. Excessive screen time may reduce opportunities for these experiences.

Attention patterns develop during formative years. Rapid transitions between images and stimuli can shape expectations about pacing. Short-form content encourages brief engagement cycles. When exposure is frequent, children may adapt to high stimulation levels.

Sleep is especially important during development. Blue light from screens can interfere with circadian rhythms. Late-night device use may reduce sleep quality. Insufficient rest affects mood, learning, and emotional regulation.

Social development also depends on real-world interaction. Play with peers teaches negotiation, cooperation, and conflict resolution. Unstructured time encourages imagination. Digital interaction can complement these experiences but should not replace them.

Excessive reliance on digital entertainment may reduce opportunities for outdoor activity. Physical movement supports motor skills and overall health. Outdoor play strengthens coordination and resilience. Screen-based environments often encourage sedentary behavior.

The content children consume matters significantly. Algorithms may recommend videos or games based on engagement patterns. Without careful supervision, exposure can include material that is not age-appropriate. Parental oversight and platform safeguards are essential.

Digital platforms frequently use engagement-driven models. Content that captures attention is promoted more widely. This structure does not inherently distinguish developmental suitability. Therefore, filtering mechanisms and parental controls become critical tools.

Children also observe adult behavior. When caregivers frequently prioritize devices over direct interaction, norms are transmitted implicitly. Modeling balanced usage demonstrates healthy boundaries. Consistency reinforces expectations.

Educational institutions face challenges integrating technology effectively. Digital tools can enhance collaboration and access to resources. However, unstructured use during instruction may reduce focus. Structured guidelines support productive implementation.

The cognitive load associated with multitasking is significant. Switching between applications while studying can reduce retention. Deep learning requires uninterrupted attention. Encouraging focused sessions improves comprehension.

Emotional regulation develops through direct interpersonal feedback. Facial expressions, tone, and physical presence help children interpret social cues. Digital communication limits these signals. Supplementing screen interaction with in-person engagement is important.

Online social environments can also influence self-image. Exposure to curated content may shape expectations about appearance and success. Children and adolescents are particularly sensitive to comparison. Parental guidance and media literacy education help mitigate effects.

The accessibility of digital devices raises concerns about dependency. When entertainment is always available, boredom tolerance may decrease. Yet boredom can stimulate creativity. Balancing stimulation with downtime is essential.

Research continues to examine the long-term effects of screen exposure. Evidence suggests moderation is key. Structured usage aligned with developmental stages supports healthier outcomes. Blanket prohibitions may be less effective than balanced integration.

Family routines can establish boundaries around device usage. Device-free meals and shared activities promote connection. Clear expectations help children understand limits. Consistency fosters self-discipline.

Technology companies can contribute by designing age-appropriate experiences. Content rating systems and parental controls enhance safety. Default settings that prioritize protection support families. Transparency about algorithms improves trust.

Children of the screen will inherit a world where digital literacy is fundamental. Understanding how algorithms function will be as important as traditional literacy. Education systems must adapt accordingly. Teaching critical engagement with technology prepares future generations.

At the same time, screen-based tools can empower creativity. Children can learn coding, design, music production, and storytelling digitally. These skills expand opportunities. The objective is not avoidance, but intentional use.

Community engagement remains vital. Sports, arts, and collaborative projects encourage holistic development. Participation in offline activities balances digital exposure. Diversity of experiences strengthens adaptability.

The neon glow of devices can captivate attention, but childhood requires varied stimuli. Sensory diversity, social interaction, and physical movement contribute to well-rounded growth. Overreliance on screens risks narrowing developmental pathways.

Policy discussions increasingly address digital well-being for minors. Guidelines on screen time, content regulation, and data protection aim to safeguard children. Collaboration between families, educators, and policymakers enhances effectiveness.

Ultimately, children of the screen will shape the future of digital culture. Their habits, expectations, and skills will influence technological evolution. Guiding their development thoughtfully ensures resilience.

Technology will remain integral to their lives. Preparing them to use it responsibly is essential. With balance, structure, and education, screens can serve as tools for growth rather than substitutes for lived experience.

By acknowledging both the benefits and risks, societies can create environments that support healthy childhood development in the digital age. Intentional design, informed parenting, and adaptive education together form the foundation for responsible integration of technology into early life.

CHAPTER . 23

Education in the Age of Distraction

Education has always been an investment in the future. It prepares individuals not only for employment, but for citizenship, critical thinking, and lifelong learning. At its best, education cultivates curiosity, discipline, and the ability to engage deeply with complex ideas. It trains attention as much as it transmits knowledge.

In the age of distraction, education faces unprecedented competition. Students now carry devices capable of delivering unlimited content. Entertainment, messaging, and social interaction are available within seconds. The classroom no longer competes only with peer conversation—it competes with global media ecosystems.

Attention is the primary resource required for learning. Without sustained focus, information cannot move from short-term exposure into long-term understanding. Deep comprehension requires uninterrupted engagement. Yet digital environments condition rapid shifts between stimuli.

Multitasking has become normalized among students. It is common for learners to alternate between coursework and social media, streaming content, or messaging applications. Research consistently shows that task-switching reduces retention and increases cognitive load. The brain must repeatedly reorient, consuming mental energy.

The architecture of digital platforms encourages interruption. Notifications appear during study sessions. Visual alerts and sounds signal new messages. Even when silenced, the presence of a device can create anticipatory distraction. The possibility of interruption affects concentration.

Short-form content further shapes expectations. Students accustomed to brief videos and rapid transitions may find extended lectures demanding. Long readings require sustained effort. The patience necessary for complex problem-solving can feel unfamiliar.

Education, however, depends on complexity. Scientific reasoning, historical analysis, and mathematical proof require layered understanding. These disciplines demand endurance in thought. If attention is fragmented, mastery becomes more difficult.

The distraction environment does not imply reduced intelligence. Rather, it suggests altered cognitive habits. Exposure to constant stimulation can recalibrate focus thresholds. Students may adapt to high-input environments and struggle with low-stimulation tasks.

Teachers increasingly integrate technology into instruction. Digital tools can enhance collaboration and access to resources. Interactive simulations, online research, and multimedia presentations enrich learning experiences. When used strategically, technology supports engagement.

However, unstructured device usage can undermine classroom focus. Without clear boundaries, devices may shift from tools to temptations. Establishing policies for use during instruction helps preserve learning integrity. Structure enables balance.

The attention crisis intersects directly with education. Students must learn not only content, but also how to manage focus. Attention management is becoming a core academic skill. Mindfulness practices and structured study techniques can strengthen this capacity.

Educational institutions can design environments that minimize unnecessary distraction. Quiet study spaces, device-free sessions, and clear guidelines reduce cognitive fragmentation. Intentional scheduling of breaks supports sustainable concentration.

Digital literacy should be embedded within curricula. Students benefit from understanding how algorithms influence information exposure. Awareness of recommendation systems encourages critical evaluation. Knowledge about media dynamics strengthens analytical skills.

Educators also face pressure to maintain engagement. Platforms accustomed to rapid stimulation may influence expectations about pacing. Lessons may need to incorporate interactive elements without sacrificing depth. Balance is key.

Assessment methods may evolve alongside technology. Open-book formats, project-based evaluations, and collaborative assignments emphasize applied understanding. These approaches reward synthesis rather than memorization alone. Technology can support these models.

Yet foundational skills remain essential. Reading comprehension, mathematical reasoning, and writing proficiency require sustained practice. These competencies form the basis for advanced learning. Protecting time for deep practice is critical.

Parental involvement contributes significantly to educational outcomes. Supporting structured homework routines, limiting excessive screen time, and encouraging reading habits reinforce academic discipline. Collaboration between families and schools strengthens consistency.

The home environment influences study habits. Quiet spaces, predictable schedules, and reduced device interference improve focus. Establishing designated study areas can help separate learning from entertainment.

Teachers can model focused engagement by minimizing personal device use during instruction. Demonstrating sustained attention reinforces norms. Consistency between expectations and behavior enhances credibility.

The age of distraction requires intentional countermeasures. Structured curriculum design, attention training, and digital boundaries collectively address challenges. Education systems must adapt without abandoning rigor.

Technology can also assist in promoting focus. Applications that block distracting sites during study periods help maintain concentration. Timers and productivity tools encourage structured work sessions. These tools transform devices into allies.

The goal is not to remove technology from education but to align it with cognitive development. When used thoughtfully, digital tools expand access and creativity. When left unchecked, they fragment attention.

Students today must navigate an environment saturated with information. Teaching them how to evaluate sources, prioritize tasks, and manage time prepares them for adulthood. These meta-skills are increasingly important.

The distraction landscape is not static. Emerging technologies will continue to influence learning environments. Continuous adaptation in pedagogy ensures relevance. Educators and policymakers must collaborate.

The neon glow of devices does not have to overshadow learning. With structured guidance, clear boundaries, and intentional integration, education can thrive. Focus can be protected even within digitally rich ecosystems.

Education in the age of distraction ultimately revolves around attention stewardship. Schools and families share responsibility for cultivating deep concentration. When attention is trained and valued, knowledge acquisition strengthens.

By acknowledging the influence of digital systems on learning, societies can respond strategically rather than reactively. Balanced policies, informed teaching practices, and conscious usage patterns together preserve the integrity of education.

In doing so, education can remain a pathway not only to employment, but to thoughtful participation in an increasingly complex world.

CHAPTER . 24

The Global Cultural Homogenization

Culture has always evolved through exchange. Trade routes, migration, and communication technologies have connected societies across distance for centuries. This exchange enriched art, language, cuisine, and philosophy. Diversity grew through interaction. Local traditions adapted while retaining distinct identities.

The digital age has accelerated cultural exchange to an unprecedented scale. Content now travels globally within seconds. Trends can emerge in one region and spread worldwide almost instantly. Platforms allow users from different continents to consume the same media simultaneously.

While this connectivity fosters shared experiences, it also contributes to cultural homogenization. Algorithms prioritize content that performs well across large audiences. Widely appealing material often receives greater visibility. As a result, certain aesthetics, narratives, and formats dominate feeds.

The logic of engagement rewards universality. Content designed to attract broad attention may overshadow localized expression. Viral trends often replicate similar structures: short formats, high emotional impact, clear hooks. These patterns can become global standards.

Language itself is influenced by digital communication. Abbreviations, memes, and platform-specific terminology spread rapidly. Expressions originating within niche communities can enter mainstream usage. While linguistic evolution is natural, algorithmic amplification accelerates convergence.

Visual culture is similarly affected. Filters, editing styles, and design templates standardize presentation. Certain color palettes, framing techniques, and video transitions become ubiquitous. Repetition across platforms can reduce aesthetic diversity.

Music and entertainment also experience consolidation. Streaming algorithms recommend popular tracks and trending artists. Exposure is influenced by engagement history. Global charts often reflect platform-driven visibility rather than purely local preferences.

Cuisine and lifestyle trends circulate through digital media. Recipes, fitness routines, fashion styles, and wellness practices spread through influencer networks. While cross-cultural exchange can be positive, algorithmic reinforcement can create uniform patterns of consumption.

Cultural homogenization does not mean the disappearance of local identities. Rather, it refers to the dominance of globally optimized formats. Content that aligns with algorithmic preferences may gain prominence over region-specific traditions.

The incentive structure of digital platforms contributes to this effect. To maximize engagement, systems prioritize content with broad appeal. Highly specific or culturally nuanced material may receive limited distribution unless it achieves viral traction.

This dynamic can shape creative industries. Artists may adapt their work to align with global trends. Musicians may adjust production styles to fit streaming algorithms. Filmmakers may design content for rapid online promotion.

The result is a feedback loop. Trends that perform well are replicated. Replication increases visibility. Increased visibility reinforces the trend. Over time, convergence intensifies.

Cultural diversity remains resilient in many communities. Local festivals, languages, and traditions continue to thrive. However, exposure to dominant digital content can influence preferences, especially among younger generations. Media consumption patterns shape identity formation.

The global reach of platforms can create shared cultural moments. International events, challenges, and campaigns connect users across borders. These shared experiences foster unity. Yet they may also standardize expression.

Standardization simplifies participation. Familiar formats reduce barriers to entry. Users can replicate trends easily. Templates and editing tools lower creative complexity. Accessibility increases engagement.

However, excessive standardization can limit experimentation. When algorithms reward predictable structures, unconventional ideas may struggle to gain attention. Novel cultural expressions may require exceptional engagement to surface.

Cultural homogenization is closely linked to commercial optimization. Platforms often prioritize scalable content. Scalability favors formats that can be easily reproduced. Unique or locally tailored experiences may receive less algorithmic support.

The phenomenon extends beyond entertainment. Educational materials, news presentation styles, and even political messaging can converge toward similar digital formats. Visual design trends influence communication strategies globally.

Despite these tendencies, digital tools also enable cultural preservation. Communities can document traditions, archive languages, and share heritage widely. Online platforms can serve as repositories for cultural knowledge. Global visibility can strengthen minority voices when intentionally amplified.

The key variable remains design and intent. Algorithms optimized solely for engagement may promote homogenization. Systems designed to prioritize diversity can counterbalance convergence. Transparency about recommendation criteria supports awareness.

Cultural exchange in the digital era can be enriching when it encourages dialogue rather than replacement. Cross-cultural collaboration can generate innovation. Hybrid art forms demonstrate creative fusion. Diversity and connectivity are not mutually exclusive.

However, the speed of digital dissemination can compress cultural cycles. Trends rise and fall rapidly. Short attention spans reduce the lifespan of creative movements. Continuous novelty replaces sustained development.

This rapid turnover can discourage long-term artistic exploration. Creators may feel pressure to produce content consistently to maintain relevance. Depth may compete with frequency. Sustainable cultural growth requires patience.

The neon infrastructure of global platforms shapes what is visible. Visibility influences perceived importance. When global algorithms highlight similar types of content, cultural landscapes may appear uniform.

Addressing homogenization requires intentional effort. Supporting local creators, promoting multilingual content, and diversifying recommendations can enhance variety. Platform design can include regional balance mechanisms.

Users also influence outcomes through their choices. Following culturally diverse accounts broadens exposure. Engaging with local content supports community vitality. Deliberate consumption patterns shape algorithmic feedback.

Cultural homogenization is not an inevitable outcome of connectivity. It emerges from specific incentive structures. When engagement metrics dominate, convergence increases. When diversity metrics are valued, plurality can flourish.

The digital world can either flatten cultural expression or amplify it. It can reduce distinction or facilitate preservation. The direction depends on governance, design, and user behavior.

Global connectivity offers extraordinary opportunities for collaboration and understanding. The challenge lies in ensuring that connection does not replace identity. Balance between shared global culture and local tradition is essential.

As we move further into the digital age, preserving cultural richness requires conscious stewardship. Encouraging variety in content exposure strengthens creativity. Recognizing algorithmic influence helps prevent unintended uniformity.

The afterworld shaped by algorithms need not erase diversity. With thoughtful systems and intentional engagement, technology can celebrate differences while enabling global dialogue. Cultural plurality can coexist with global connection when design prioritizes balance.

Understanding homogenization allows societies to safeguard heritage while embracing innovation. Awareness transforms passive convergence into active cultural evolution.

CHAPTER . 25

Naming the Gods

Every transformation begins with recognition. Before a problem can be solved, it must be clearly identified. Before systems can be changed, they must be understood. Naming is not symbolic alone —it is strategic. When we name forces shaping our lives, we move from unconscious participation to conscious engagement.

The "neon gods" described throughout this book are not supernatural beings. They are systems. They are algorithms, incentive structures, design philosophies, and economic models. They are architectures that shape behavior through visibility, reward, and friction. They influence attention, perception, identity, and power.

For much of the digital era, these systems operated invisibly. Users experienced interfaces without seeing underlying logic. Feeds appeared natural. Recommendations felt intuitive. Engagement seemed self-directed. The architecture remained largely hidden behind smooth design.

Naming these systems clarifies their role. Algorithms are not neutral observers; they are decision-making frameworks coded by human priorities. Engagement metrics are not value judgments; they are business indicators. Data pipelines are not accidental flows; they are structured extraction processes.

When systems remain unnamed, they appear inevitable. Inevitability discourages reform. If something feels like nature, it resists critique. By contrast, once structures are identified, they become subject to evaluation and redesign.

Naming also reduces psychological mystification. Many users feel that digital platforms possess independent agency. In reality, these platforms reflect cumulative human decisions embedded in code. Understanding this distinction restores perspective.

The process of naming requires intellectual honesty. It requires acknowledging incentives that drive behavior. It requires recognizing how engagement economics influences content prioritization. It requires accepting that convenience can coexist with unintended consequences.

Naming the gods means identifying what is being optimized. Is attention the primary metric? Is growth prioritized over well-being? Are profits weighted above public health? Clarity about objectives reveals structural direction.

This clarity is not about assigning blame to individuals alone. Engineers, designers, investors, and users operate within incentive systems. Behavior aligns with reward structures. When rewards emphasize engagement, systems will maximize engagement.

Once these dynamics are visible, alternative designs become imaginable. If well-being were measured alongside engagement, priorities might shift. If transparency were rewarded, opacity might decrease. If diversity were valued, homogenization could slow.

Naming also empowers collective dialogue. Public conversation about digital systems becomes more productive when terminology is shared. Discussions about algorithmic influence, data ownership, and platform governance rely on common understanding.

Without naming, debates remain abstract. Vague concerns lack traction. Concrete vocabulary enables policy development. It enables education. It enables reform.

In addition, naming shifts internal habits. When users recognize that infinite scroll is engineered to prolong engagement, they can respond intentionally. When notifications are understood as behavioral triggers, they can be managed deliberately. Awareness transforms reflex into choice.

The act of naming is therefore foundational to reclaiming agency. It interrupts automatic acceptance. It introduces critical distance between stimulus and response. That distance creates freedom.

Throughout history, societies have named forces to better understand them. Economic systems, political ideologies, and scientific principles all became clearer through articulation. Language organizes complexity. Precision reduces confusion.

The neon systems of the digital age require similar articulation. Attention extraction, behavioral prediction, engagement optimization, and data monetization are not abstract trends—they are describable mechanisms. Naming them does not reject technology; it clarifies its operation.

Clarity allows constructive reform rather than reactive resistance. When systems are misunderstood, responses can become extreme. When systems are understood, solutions can be targeted. Thoughtful change depends on accurate diagnosis.

Naming also supports education. Students who understand algorithmic logic can interpret feeds critically. Professionals who grasp incentive structures can design ethically. Citizens who comprehend data economics can participate in governance discussions effectively.

The gods of the digital afterworld thrive in ambiguity. Ambiguity sustains passivity. When users feel overwhelmed or uncertain about how systems work, disengagement or resignation may follow. Naming reduces that uncertainty.

This chapter marks a transition from analysis to action. Earlier chapters examined consequences: attention crises, burnout, surveillance, polarization, and identity performance. Naming connects those effects to their structural sources.

By identifying the systems involved, we prepare to reclaim core human capacities—attention, community, creativity, and autonomy. Recognition precedes redesign. Understanding precedes transformation.

The neon glow is not destiny. It is infrastructure. Infrastructure can be rebuilt. Incentives can be adjusted. Design can evolve.

Naming the gods is therefore the first step toward reshaping the afterworld. Once we can clearly describe what influences us, we can decide how to respond. Language becomes leverage.

From here, the book moves from diagnosis to restoration. With clarity established, the next chapters will explore how attention can be reclaimed, how literacy can evolve, and how communities can rebuild resilience within digital systems.

Awareness is not the end of the journey—it is the beginning of empowerment.

CHAPTER . 26

Reclaiming Attention as Sacred

Attention is the gateway to experience. It determines what we notice, what we value, and what we remember. In many traditions, attention has been treated as sacred because it shapes consciousness itself. What we consistently focus on becomes the architecture of our inner life. Reclaiming attention means reclaiming agency.

In the digital age, attention has been externalized and monetized. Platforms compete to capture focus and retain it as long as possible. Notifications, infinite feeds, autoplay features, and personalized recommendations are designed to reduce friction. These systems optimize for engagement, not necessarily for well-being.

When attention is constantly directed outward, inner awareness weakens. Reflection requires stillness. Insight requires sustained concentration. Creativity requires uninterrupted mental space. Without intentional boundaries, external stimuli dominate internal clarity.

Reclaiming attention begins with recognition. We must become aware of how frequently our focus is redirected. Each interruption fragments thought. Each alert shifts cognitive state. Awareness transforms interruption from automatic response into conscious choice.

Sacred attention does not mean withdrawal from technology. It means deliberate engagement. It means deciding when and why to connect. It means distinguishing between meaningful interaction and habitual scrolling. Intention replaces reflex.

One practical step is establishing boundaries around devices. Time blocks for focused work create structured environments for concentration. Turning off nonessential notifications reduces involuntary shifts in attention. Physical separation from devices during deep tasks strengthens focus.

Another practice involves cultivating single-tasking. Completing one activity before beginning another enhances cognitive depth. Multitasking may feel productive, but it often reduces quality and increases fatigue. Sustained engagement builds mental endurance.

Mindfulness training offers additional support. By observing thoughts without immediately reacting to them, individuals strengthen attentional control. Meditation practices develop awareness of internal states. Over time, this enhances the ability to resist distraction.

Digital systems will continue to evolve, so attention stewardship must be ongoing. Periodic evaluation of usage patterns helps maintain balance. Reflecting on screen time, content sources, and emotional responses reveals areas for adjustment.

Reclaiming attention also involves redesigning environments. Workspaces free of unnecessary alerts promote concentration. Study areas with minimal visual clutter reduce cognitive load. Environments shape behavior, so intentional design supports focus.

Communities can reinforce attentional values collectively. Shared device-free gatherings normalize presence. Group commitments to focused meetings encourage engagement. Cultural norms influence individual habits significantly.

Educational institutions can teach attention management as a core skill. Students benefit from learning how distraction affects cognition. Training in deep reading, structured note-taking, and deliberate practice strengthens learning capacity. Attention is foundational to academic success.

Reclaiming attention also restores emotional balance. Constant stimulation can heighten stress responses. Periods of quiet allow the nervous system to stabilize. Calm states support resilience and clarity.

The sacredness of attention lies in its scarcity and power. Where attention goes, energy flows. When attention is fragmented, purpose becomes diluted. When attention is focused, action gains direction.

Digital platforms often rely on habitual engagement. Breaking automatic cycles reduces dependence. Conscious pauses before opening applications interrupt momentum. Small moments of reflection can prevent extended distraction.

Reclaiming attention can be framed as an act of self-respect. Protecting focus honors mental capacity. It acknowledges that cognitive resources are finite. It affirms the value of depth over constant stimulation.

This process requires patience. Distraction habits form over time and may not change instantly. Gradual reduction of unnecessary engagement supports sustainable improvement. Incremental change is more effective than abrupt extremes.

Technology itself can assist in attention stewardship. Focus modes, application timers, and screen time reports provide feedback. These tools transform devices from sources of distraction into instruments of awareness. Design can align with intention.

Reclaiming attention also strengthens creativity. Deep concentration enables synthesis of complex ideas. Innovation often emerges from extended problem-solving sessions. Sustained focus allows patterns to surface naturally.

Social relationships benefit as well. Undivided presence during conversation enhances empathy. Listening without interruption signals respect. Attention is a form of care.

The neon systems of distraction depend on unconscious participation. When users become aware of attentional dynamics, they regain control. Awareness introduces choice. Choice restores autonomy.

Sacred attention does not require perfection. Occasional distraction is natural. The goal is balance, not elimination. Consistency over time shapes outcomes.

By treating attention as valuable, individuals can prioritize activities that align with long-term goals. Intentional engagement replaces impulsive consumption. Reflection precedes reaction.

In reclaiming attention, we begin reshaping the broader digital environment. Collective focus on well-being can influence design standards. Consumer demand can shift market incentives. Cultural priorities evolve when awareness spreads.

Attention is the foundation of consciousness. To protect it is to protect thought itself. In the age of algorithms, guarding attention is essential for independence and growth.

Reclaiming attention as sacred is not a rejection of technology. It is a commitment to intentional living within technological systems. By setting boundaries, practicing focus, and valuing depth, we transform distraction into discipline.

This chapter marks the transition from understanding systems to actively reshaping engagement. With attention restored, we can now explore how literacy, community, and design can further support resilience in the digital age.

CHAPTER . 27

Digital Literacy as Civic Armor

In previous generations, civic strength depended on literacy in language, law, and public institutions. Today, it also depends on literacy in digital systems. The ability to navigate online environments critically is no longer optional. It is essential for informed participation in society.

Digital literacy extends beyond knowing how to use devices. It includes understanding how algorithms curate information, how data is collected, and how engagement metrics shape visibility. It requires awareness of incentives embedded in platforms. It transforms users from passive consumers into informed participants.

Without digital literacy, individuals are vulnerable to manipulation. Misinformation can spread rapidly through persuasive design and emotional framing. Deepfakes, targeted advertisements, and algorithmic amplification can distort perception. Awareness reduces susceptibility.

Civic armor is a metaphor for resilience. Just as physical armor protects the body, digital literacy protects judgment. It does not eliminate exposure to information. Instead, it strengthens the ability to evaluate it critically.

One core component of digital literacy is understanding algorithmic influence. Platforms rank content based on engagement signals. These signals include clicks, shares, watch time, and reactions. Popularity does not always equal accuracy or importance.

Recognizing this distinction helps users interpret feeds more effectively. When content appears prominently, it reflects system optimization rather than inherent value. Understanding this mechanism prevents overreliance on visibility as a measure of truth.

Another component is media verification. Cross-checking information across multiple reputable sources reduces the risk of accepting false claims. Fact-checking organizations provide additional context. Corroboration strengthens confidence in conclusions.

Privacy awareness is also central to civic protection. Knowing how data is collected and used enables informed consent. Adjusting privacy settings, reviewing permissions, and limiting unnecessary sharing reduces exposure. Control over personal information enhances autonomy.

Digital literacy includes recognizing persuasive design techniques. Features such as infinite scroll, autoplay, and notification systems are intentionally structured to sustain engagement. Awareness of these mechanisms supports intentional use rather than automatic interaction.

Critical thinking skills are foundational. Evaluating evidence, identifying logical fallacies, and distinguishing opinion from fact are transferable abilities. These skills apply both online and offline. Strengthening them increases resilience against manipulation.

Educational systems play a key role in cultivating digital literacy. Curricula can incorporate instruction on algorithmic systems, data ethics, and online safety. Teaching students how platforms function equips them for modern civic life. Knowledge builds confidence.

Community programs can extend digital literacy beyond schools. Public workshops, library initiatives, and online courses can support adults as well. Lifelong learning ensures that citizens adapt alongside technology. Inclusivity enhances societal strength.

Platform transparency contributes to civic armor. Clear explanations of recommendation criteria reduce uncertainty. Disclosure about sponsored content, political advertising, and data usage improves trust. Transparency fosters accountability.

Users can practice digital hygiene. Limiting exposure to sensational content reduces emotional volatility. Following diverse and credible sources broadens perspective. Conscious curation strengthens informational balance.

Recognizing emotional manipulation is another protective skill. Content designed to provoke outrage, fear, or urgency often aims to maximize engagement. Pausing before sharing emotionally charged material interrupts amplification cycles. Reflection reduces unintended spread.

Digital literacy also involves understanding synthetic media. Awareness that images, audio, and video can be generated artificially encourages verification. Skepticism becomes calibrated rather than absolute. Balanced trust supports social stability.

Civic engagement in the digital age requires awareness of data-driven influence. Targeted messaging can shape perceptions during political campaigns. Recognizing personalization reduces the illusion of universal messaging. Informed citizens evaluate claims independently.

Community dialogue strengthens literacy efforts. Discussing online experiences with peers promotes shared learning. Collective awareness enhances resilience. Conversations about technology normalize critical reflection.

Civic armor does not imply hostility toward innovation. Instead, it acknowledges complexity. Technology can empower creativity, communication, and collaboration. Literacy ensures that benefits are maximized while risks are managed.

Developing digital literacy requires practice. Engaging with varied content, questioning sources, and analyzing platform behavior builds familiarity. Experience deepens understanding. Repetition reinforces skill.

Organizations can support transparency initiatives. Publishing algorithmic guidelines and impact assessments increases public insight. Ethical standards for data governance further enhance protection. Collaboration between sectors is necessary.

Digital literacy empowers individuals to participate confidently in online discourse. It reduces fear and confusion. It increases informed debate. It strengthens democratic engagement.

The neon systems of influence are sophisticated. But informed users can navigate them effectively. Knowledge transforms complexity into manageable structure. Understanding reduces vulnerability.

Civic armor ultimately consists of awareness, critical thinking, privacy management, and emotional regulation. These elements together create resilience. They enable engagement without exploitation.

In an interconnected world, literacy is power. When citizens understand the mechanisms shaping their information environment, they can make deliberate choices. Digital literacy ensures that participation remains grounded in agency.

By strengthening these skills, societies prepare themselves for ongoing technological evolution. The goal is not resistance to change, but readiness for it. With civic armor in place, communities can interact with digital systems confidently and responsibly.

CHAPTER . 28

Rebuilding Community Offline

Human beings are not designed to live in isolation. We evolved in close-knit groups where cooperation ensured survival and shared responsibility sustained progress. Community was not optional—it was foundational. Belonging provided security, identity, and mutual accountability.

Digital networks have expanded the number of connections we can maintain, but they have not replaced the depth of physical community. Online interaction can supplement relationships, yet embodied presence creates a different quality of bond. Face-to-face communication engages tone, expression, posture, and shared environment in ways screens cannot fully replicate.

Rebuilding community offline begins with intentional proximity. Physical gatherings restore sensory richness to interaction. Shared meals, conversations, and collaborative activities cultivate trust. Repetition of in-person contact strengthens familiarity and reliability.

Local engagement forms the backbone of resilient societies. Neighborhood relationships create informal support systems. When people know those around them, cooperation becomes easier. Mutual recognition reduces anonymity and increases accountability.

Community spaces play an essential role in this process. Parks, libraries, schools, faith centers, and community halls provide environments for gathering. These spaces encourage interaction beyond digital mediation. Accessible public venues foster inclusivity.

Rebuilding offline community requires time investment. Unlike digital engagement, which offers instant interaction, physical connection develops gradually. Trust grows through repeated experiences. Consistency matters more than intensity.

Shared projects are particularly effective at strengthening bonds. Volunteer initiatives, local improvement efforts, and group learning activities create common goals. Working toward tangible outcomes reinforces cooperation. Collective effort deepens relationships.

Intergenerational interaction also enhances community health. When younger and older members engage directly, knowledge transfers naturally. Experience and innovation complement each other. Physical proximity encourages mentorship and learning.

Reducing excessive screen dependence supports offline connection. Device-free gatherings encourage full presence. Establishing boundaries around technology during social events enhances engagement. Attention directed toward people strengthens relationships.

Community resilience improves when social ties are strong. During crises, neighbors can provide assistance. Information spreads more effectively through trusted local networks. Shared understanding fosters coordinated response.

Offline relationships also support mental well-being. Regular in-person interaction reduces feelings of isolation. Physical presence can regulate stress and improve emotional balance. Human contact remains a core psychological need.

Cultural traditions often serve as anchors for community identity. Festivals, local ceremonies, and collective celebrations reinforce shared values. Participating in these events strengthens continuity. Tradition and community reinforce each other.

Urban environments can present challenges for offline community building. High population density does not automatically produce connection. Intentional design of shared spaces is necessary. Mixed-use areas and pedestrian-friendly environments encourage interaction.

Rural communities may also face challenges, particularly when digital engagement replaces local gatherings. Even in smaller populations, virtual interaction can reduce physical meeting frequency. Maintaining local events ensures continuity.

Schools can serve as hubs for community engagement. Parent-teacher gatherings, public lectures, and extracurricular activities create shared experiences. Educational institutions connect families and residents. They anchor community networks.

Workplaces can contribute as well. Team-building activities, collaborative projects, and in-person meetings foster interpersonal trust. Organizational culture benefits from direct communication. Face-to-face interaction strengthens cohesion.

Community rebuilding also involves listening. Open forums and town meetings encourage dialogue. When individuals feel heard, engagement increases. Inclusive communication builds mutual respect.

Trust grows when interactions are consistent and transparent. Reliability in small commitments accumulates over time. Shared experiences create a sense of belonging. Belonging is the foundation of resilience.

Offline community does not require rejection of digital tools. Technology can facilitate coordination and event planning. Online platforms can organize in-person meetings. The goal is integration, not replacement.

Intentional scheduling helps maintain balance. Setting regular times for community activities ensures continuity. Treating social engagement as a priority rather than an afterthought reinforces commitment. Structure supports participation.

Physical presence also enhances empathy. Observing body language and tone allows deeper understanding. Misinterpretations are less likely when communication is direct. Nuance emerges more clearly in shared space.

Community health depends on inclusivity. Welcoming diverse participants strengthens networks. Encouraging participation across backgrounds broadens perspectives. Diversity enriches collective experience.

Rebuilding offline connections requires patience. Trust cannot be manufactured quickly. It develops through repeated positive interactions. Consistency over time produces stability.

Local economic engagement further supports community strength. Supporting neighborhood businesses keeps resources circulating within the area. Shared economic activity fosters interdependence. Economic ties reinforce social ones.

Volunteerism is another pathway to connection. Contributing time and skills to local causes builds purpose. Shared responsibility strengthens bonds. Service aligns individual effort with collective benefit.

In a world increasingly mediated by screens, physical gatherings regain importance. The simple act of meeting in person counters fragmentation. It restores the sensory depth of interaction. Embodied presence rebalances digital intensity.

Rebuilding community offline enhances resilience against many challenges discussed earlier in this book. Strong communities buffer against loneliness, burnout, and polarization. They provide support structures independent of algorithmic systems. They ground identity in shared reality.

The neon glow of digital life can feel expansive, but community thrives in shared physical environments. By investing in local relationships, societies strengthen cohesion. Connection becomes tangible rather than abstract.

Rebuilding offline community is not nostalgic—it is strategic. It ensures that technological progress does not erode fundamental human bonds. When digital tools complement rather than replace physical interaction, balance is restored.

Strong communities form the foundation for democratic participation, psychological well-being, and cultural vitality. By prioritizing in-person engagement, societies create durable networks of trust. These networks support stability in times of change.

Chapter . 29

Ethical Technology Design

Technology is not neutral in its impact. While tools themselves can be used for many purposes, the way they are designed strongly influences how they are used. Design choices shape behavior. Interfaces, defaults, incentives, and architecture guide user decisions, often without explicit awareness.

Ethical technology design begins with intention. It asks not only what a system can do, but what it should do. It evaluates consequences alongside capabilities. It prioritizes human well-being as a central metric rather than treating engagement as the sole indicator of success.

In many digital platforms, engagement has become the dominant measure of performance. Time spent, clicks, and interactions are tracked continuously. These metrics drive optimization. When engagement is equated with value, design decisions follow accordingly.

Ethical design expands the definition of success. It may include user satisfaction, mental health outcomes, transparency, accessibility, and long-term sustainability. It balances short-term activity with long-term impact. It recognizes that maximizing attention is not always aligned with maximizing benefit.

One principle of ethical design is transparency. Users should understand how systems operate. Clear explanations of algorithms, data usage, and recommendation logic reduce confusion. Transparency builds trust and accountability.

Another principle is consent. Data collection and feature usage should be optional and clearly explained. Users should have meaningful control over their information. Consent must be informed, not merely procedural.

Privacy by design is a related concept. Systems can minimize data collection by default. Only necessary information should be gathered. Reducing stored data lowers risk and increases security. Simplicity strengthens protection.

Ethical design also considers friction. Not all friction is harmful. In fact, small pauses can encourage reflection. Confirmation steps before sharing sensitive content can prevent impulsive dissemination. Thoughtful friction can reduce unintended consequences.

Attention-aware design is increasingly important. Features such as infinite scroll and autoplay remove natural stopping points. Reintroducing deliberate endpoints can help users disengage intentionally. Design can support balanced usage patterns.

Time-awareness tools contribute to ethical outcomes. Screen-time summaries, usage reminders, and break prompts increase self-awareness. When users can see their habits clearly, they can adjust accordingly. Feedback promotes autonomy.

Inclusivity is another core element. Technology should accommodate diverse abilities, languages, and contexts. Accessibility features ensure broader participation. Ethical design recognizes that systems serve varied populations.

Safety mechanisms are equally important. Moderation tools, reporting systems, and content filters help reduce harm. Clear community guidelines establish expectations. Consistency in enforcement maintains credibility.

Ethical design must also address persuasive techniques. Platforms often use psychological principles to sustain engagement. While motivation can be beneficial, excessive manipulation undermines autonomy. Transparency about design intent protects users.

Long-term thinking differentiates ethical systems from purely extractive ones. Instead of optimizing only for immediate growth, ethical platforms evaluate sustainability. They consider how design choices affect users over years, not minutes.

Collaboration between technologists, ethicists, policymakers, and users strengthens outcomes. Diverse perspectives help identify unintended effects. Cross-disciplinary input improves decision-making. Shared responsibility enhances accountability.

Open standards and interoperability contribute to ethical ecosystems. When users can move data between platforms easily, dependence decreases. Competition improves when switching costs are low. Flexibility supports autonomy.

Ethical design also encourages moderation of addictive features. Limiting variable reward mechanisms, reducing notification overload, and providing customizable controls help maintain balance. Users should have the ability to tailor experiences to their preferences.

Education complements design reform. Users benefit from understanding how interfaces influence behavior. Awareness enables resistance to unwanted patterns. Knowledge transforms interaction from passive to active.

Companies that adopt ethical design principles may build stronger long-term trust. Transparency and user-centered development can enhance reputation. Sustainable engagement often outperforms short-term extraction strategies.

Regulation may support ethical standards when market incentives alone are insufficient. Policy frameworks can establish minimum requirements for privacy, accessibility, and transparency. Governance helps align innovation with societal values.

Ethical technology design recognizes that systems shape habits, attention, and relationships. Therefore, responsibility extends beyond functionality. It includes impact assessment and continuous evaluation.

Iterative improvement is essential. As technology evolves, ethical guidelines must adapt. Feedback loops between users and developers ensure responsiveness. Ongoing assessment prevents stagnation.

Designers can incorporate well-being metrics alongside engagement metrics. Surveys, opt-in feedback, and research partnerships can inform development. Data can be used to enhance health rather than solely increase retention.

The goal of ethical design is alignment between human flourishing and technological capability. It seeks to ensure that tools amplify autonomy, creativity, and connection. It reframes innovation as service rather than extraction.

In the context of the broader themes of this book, ethical design addresses many of the systems previously described—attention capture, surveillance models, algorithmic polarization, and performance incentives. By adjusting design principles, these forces can be moderated.

Ethical technology does not require rejecting digital progress. Instead, it requires intentional architecture. It calls for systems that respect users as participants rather than resources. It emphasizes dignity, transparency, and sustainability.

The neon infrastructure of the digital world can either prioritize engagement above all else or balance engagement with well-being.

Ethical design chooses the latter. It integrates human values into code and interface decisions.

By embedding ethical principles into development processes, societies can shape digital environments proactively. Technology then becomes a tool that aligns with collective aspirations. Responsibility becomes a shared foundation for innovation.

Chapter . 30

Regulating the Invisible Hand

Markets have long relied on the idea of the "invisible hand," the belief that decentralized decisions can produce collective benefit. Competition, consumer choice, and innovation often generate efficiency and growth. When functioning well, markets reward value creation and discourage stagnation.

In the digital era, however, the invisible hand operates within algorithmic systems that were not present in earlier economic models. Platforms shape visibility, influence consumer behavior, and mediate communication at global scale. These systems are not merely marketplaces; they are infrastructure. Their design affects how markets function.

Regulation in this context does not mean suppressing innovation. It means establishing guardrails that ensure fairness, transparency, and accountability. Rules help align private incentives with public interest. Without oversight, concentrated power can distort outcomes.

Digital markets exhibit characteristics that differ from traditional industries. Network effects, data accumulation, and platform dependency can reduce competition naturally. Once a platform reaches critical mass, switching costs increase. Users, businesses, and creators may depend on access to its ecosystem.

Regulation can address these structural dynamics. Antitrust frameworks may prevent excessive consolidation. Data portability requirements can reduce lock-in effects. Interoperability standards can enable competition by allowing systems to communicate across platforms.

Transparency is another regulatory priority. Clear disclosure of algorithmic criteria helps users understand how content is prioritized. Public reporting on data practices enhances accountability. When systems are more visible, trust improves.

Political advertising regulation is also relevant. Digital campaigns can use precise targeting based on behavioral data. Oversight ensures that such practices do not undermine democratic processes. Disclosure requirements help voters understand who is funding messages and why they are seeing them.

Regulation can also protect privacy. Laws may limit unnecessary data collection and require secure storage practices. Consent mechanisms should be clear and meaningful. Users should retain control over their information.

Economic concentration in digital markets has prompted renewed attention to competition policy. Regulators may evaluate mergers and acquisitions that could reduce market diversity. Maintaining competitive ecosystems encourages innovation and consumer choice.

However, regulation must balance flexibility with structure. Overly rigid rules can stifle experimentation. Excessive bureaucracy may slow beneficial development. Effective policy adapts to technological change while preserving core principles.

International coordination is increasingly important. Digital platforms operate globally, often across multiple jurisdictions. Harmonizing standards can reduce regulatory gaps. Cooperation between governments strengthens enforcement.

Public consultation contributes to balanced governance. Engaging stakeholders—including users, developers, businesses, and civil society—ensures diverse perspectives. Inclusive dialogue improves legitimacy and effectiveness.

Regulation also intersects with labor markets. As automation and platform-based work expand, policies may address worker protections, benefits, and rights. Clear guidelines can ensure fair treatment within digital ecosystems.

Tax frameworks may evolve to reflect digital economic structures. Ensuring equitable contribution across industries supports public services. Revenue generated from digital commerce can contribute to infrastructure, education, and healthcare.

The invisible hand functions best when markets are competitive and transparent. Regulation does not replace market mechanisms; it shapes their environment. Guardrails create stability without eliminating freedom.

In digital systems, governance can extend beyond government alone. Industry standards, independent audits, and ethical certifications contribute to oversight. Multi-stakeholder models distribute responsibility.

Accountability mechanisms strengthen public confidence. Reporting requirements, impact assessments, and external reviews ensure compliance. Measurable benchmarks allow progress evaluation.

Regulation can also address algorithmic fairness. Systems that influence credit, employment, housing, or public visibility should avoid discriminatory bias. Independent testing can detect unintended disparities. Oversight reduces systemic inequity.

Innovation thrives when rules are clear. Predictability encourages investment. Ambiguity increases risk. Well-defined frameworks support responsible development.

The goal of regulating the invisible hand is not control for its own sake. It is alignment between technological capability and societal values. Regulation can guide systems toward transparency, fairness, and sustainability.

Digital infrastructure now plays a central role in economic life. Therefore, oversight of that infrastructure affects broad outcomes. Ensuring stability in digital markets protects consumers, workers, and innovators alike.

In the context of this book, regulation serves as a counterbalance to surveillance-based models, wealth concentration, and engagement-driven design. By setting boundaries, societies can mitigate unintended consequences. Structured governance supports resilience.

The neon systems that shape modern life are powerful, but they are not beyond influence. Through thoughtful regulation, communities can align market incentives with public good. Transparency, competition, privacy, and fairness form the foundation.

Effective regulation evolves alongside technology. Continuous review ensures relevance. Collaboration between policymakers and technical experts improves precision.

When properly designed, regulation enhances trust in digital systems. Trust encourages participation. Participation sustains innovation. Stability and growth can coexist.

Regulating the invisible hand does not halt progress. It ensures that progress remains beneficial and inclusive. By embedding accountability into digital markets, societies can create environments where innovation serves humanity rather than overriding it.

CHAPTER . 31

Redesigning Education for Depth

Education in the digital age must do more than deliver information. Information is abundant and instantly accessible. What is scarce is the ability to process, evaluate, synthesize, and apply it with wisdom. Depth of understanding is now more valuable than mere exposure to content.

Redesigning education for depth begins with restoring sustained attention as a core objective. Learning environments should prioritize uninterrupted engagement with complex material. Extended reading, problem-solving sessions, and structured discussion cultivate intellectual endurance. Depth develops through time, not speed.

Curricula must balance foundational knowledge with critical thinking. Students need essential facts, but they also need frameworks for interpretation. Analytical skills enable learners to assess claims, identify bias, and construct reasoned arguments. Education should emphasize reasoning alongside recall.

Project-based learning supports depth by encouraging integration of concepts. When students apply knowledge to real-world problems, understanding strengthens. Research, collaboration, and presentation develop synthesis skills. Learning becomes active rather than passive.

Assessment methods can reinforce depth-oriented goals. Instead of relying solely on short-answer formats, evaluations can include essays, research projects, and oral examinations. These formats require elaboration and reflection. They reward comprehension over memorization.

Digital tools can support deeper learning when used intentionally. Interactive simulations, collaborative platforms, and research databases expand access to resources. However, technology should enhance instruction rather than distract from it. Structured integration is essential.

Attention management should be explicitly taught within education systems. Students benefit from learning how distraction affects cognition. Training in focus techniques, time-blocking, and study routines improves academic outcomes. Meta-learning strengthens independence.

Redesigning education also involves pacing. Slower, deliberate instruction allows concepts to consolidate. Revisiting material through spaced repetition enhances retention. Depth grows through reinforcement and reflection.

Classroom environments can be structured to minimize unnecessary interruptions. Device policies, quiet study periods, and designated discussion times create stability. Predictable routines support concentration. Consistency fosters engagement.

Interdisciplinary learning contributes to depth by connecting subjects. Real-world challenges rarely fit into isolated categories. Integrating science, humanities, and social studies promotes holistic understanding. Cross-disciplinary thinking mirrors complex reality.

Educators can encourage curiosity-driven exploration. When students pursue questions that interest them, intrinsic motivation increases. Inquiry-based learning fosters ownership of knowledge. Engagement becomes internally driven rather than externally imposed.

Critical media literacy should be embedded within the curriculum. Students must understand how digital systems shape information exposure. Learning about algorithms, data collection, and persuasive design prepares them for informed participation. Awareness strengthens discernment.

Writing and discussion are powerful tools for deep learning. Articulating ideas in structured formats clarifies thinking. Debate encourages evidence-based reasoning. Dialogue refines perspectives through exchange.

Memory development benefits from active recall practices. Testing oneself on material enhances retention. Spaced repetition strengthens neural pathways. Depth emerges through structured review.

Educators can cultivate intellectual resilience by normalizing challenge. Difficult tasks promote growth. Encouraging perseverance builds confidence in problem-solving. Struggle, when supported, becomes productive.

Community involvement enriches educational depth. Partnerships between schools, families, and local organizations provide real-world context. Service learning connects theory to practice. Experience reinforces conceptual understanding.

Redesigning education for depth also includes teacher support. Professional development enables educators to adapt to evolving environments. Training in digital integration and attention strategies enhances effectiveness. Empowered teachers strengthen institutions.

The goal is not to eliminate technology from education, but to align it with cognitive development. Devices can provide access to vast information. Structured guidance ensures that access translates into understanding. Intentional use preserves focus.

Depth requires time for reflection. Reflection consolidates learning and encourages synthesis. Journaling, discussion, and project review sessions support this process. Thinking about thinking improves comprehension.

Educational redesign should consider well-being alongside achievement. Balanced workloads prevent burnout. Sustainable pacing maintains motivation. Depth thrives in environments that respect human limits.

Assessment systems can emphasize growth over competition. Measuring progress relative to individual development encourages continuous improvement. Collaboration rather than excessive rivalry fosters shared success. Learning becomes a collective endeavor.

Institutions can also promote creativity as a central objective. Innovation depends on imagination and experimentation. Providing opportunities for artistic expression and design thinking enriches cognitive diversity. Creative skills complement analytical abilities.

The afterworld shaped by distraction demands educational environments that counterbalance fragmentation. Schools can serve as sanctuaries of focus. Structured spaces dedicated to deep engagement protect attention from external noise.

Redesigning education for depth prepares students for complex global challenges. Critical thinking, collaboration, and adaptability are essential in an interconnected world. These capabilities arise from sustained intellectual effort.

Policy support may be necessary to implement systemic change. Funding for teacher training, curriculum development, and infrastructure upgrades ensures equitable access. Investment in education yields long-term societal benefits.

By prioritizing depth, educational systems cultivate citizens capable of thoughtful participation. Knowledge combined with analytical skill strengthens democracy, innovation, and cultural resilience. Education becomes a stabilizing force in a rapidly changing environment.

Redesigning for depth is ultimately about reclaiming intellectual space. When learners are given time, structure, and support, understanding flourishes. Depth transforms information into wisdom. And wisdom prepares society for sustainable progress.

CHAPTER . 32

Parenting Beyond the Screen

Parenting has always required adaptation to the surrounding culture. Each generation of parents faces new technologies, social norms, and economic pressures. In the digital age, one of the most significant challenges is guiding children through environments saturated with screens. Technology is woven into daily life, making complete avoidance unrealistic. The goal is not elimination, but intentional balance.

Parenting beyond the screen begins with modeling behavior. Children learn more from observation than instruction. When caregivers demonstrate mindful technology use, boundaries become visible. If adults constantly prioritize devices over presence, children internalize that pattern. Consistency between guidance and example is essential.

Establishing clear household guidelines creates structure. Defined times for device use reduce ambiguity. Device-free meals and family gatherings reinforce connection. Predictable rules help children understand expectations. Structure fosters self-discipline.

Creating shared offline routines strengthens bonds. Activities such as reading together, outdoor play, cooking, or collaborative projects provide meaningful alternatives to screen time. Shared experiences build memories. These moments contribute to emotional security.

Communication about technology is equally important. Open discussions about why limits exist encourage understanding rather than resistance. Explaining how algorithms work and how attention is shaped empowers children with awareness. Transparency builds trust.

Digital literacy should begin early but remain age-appropriate. Teaching children how to evaluate online information prepares them for responsible engagement. Guidance on privacy, respectful communication, and critical thinking equips them for independence. Education reduces vulnerability.

Sleep protection is a key component of screen-aware parenting. Limiting device use before bedtime supports healthy rest. Consistent sleep routines enhance mood, focus, and development. Rest strengthens resilience.

Encouraging physical activity balances sedentary tendencies. Outdoor play, sports, and movement-based games contribute to cognitive and physical growth. Exercise improves attention and emotional regulation. Active habits counteract prolonged screen exposure.

Parental controls and content filters can provide additional safeguards. These tools help manage exposure to inappropriate material. However, controls should complement dialogue rather than replace it. Supervision combined with conversation is most effective.

Designated device-free spaces in the home reinforce boundaries. Bedrooms and dining areas can remain screen-free zones. Physical separation reduces temptation. Clear spatial rules simplify expectations.

Encouraging creativity supports development beyond consumption. Art, music, writing, and hands-on projects foster imagination. Creative expression builds confidence and problem-solving skills. These activities do not depend on algorithms for validation.

Community involvement enhances parenting efforts. Engaging with schools, neighborhood groups, and extracurricular activities provides broader social context. Shared responsibility strengthens support networks. Parenting is enriched by collaboration.

Emotional development benefits from direct interpersonal interaction. Conversations about feelings, challenges, and experiences help children build empathy. Active listening models respectful communication. Presence during dialogue reinforces connection.

Limiting passive content consumption is particularly important. While some digital resources are educational, excessive entertainment-based usage can reduce engagement in other activities. Balanced scheduling ensures diversity of experiences.

Teaching children about persuasive design techniques increases awareness. Understanding how notifications and reward systems function reduces susceptibility to compulsive behavior. Knowledge empowers self-regulation. Awareness transforms impulse into choice.

Parents can establish gradual transitions rather than abrupt restrictions. Incremental adjustments are often more sustainable. Open negotiation can reduce conflict. Flexibility within boundaries supports cooperation.

Technology-free family rituals create lasting traditions. Regular game nights, outdoor excursions, or collaborative projects strengthen identity. Rituals provide stability in an evolving digital landscape. Consistency fosters belonging.

Encouraging independent problem-solving develops confidence. When children are allowed to attempt challenges before seeking digital solutions, resilience grows. Struggle within safe boundaries builds competence. Effort cultivates mastery.

Monitoring online interaction is necessary for safety. Awareness of social media use, messaging platforms, and gaming communities helps protect children from potential risks. However, monitoring should respect privacy in developmentally appropriate ways. Balance between trust and oversight is key.

Parent education programs can support families navigating digital environments. Workshops and community resources provide guidance. Shared learning experiences strengthen parenting strategies. Collaboration enhances effectiveness.

Parenting beyond the screen also includes fostering offline friendships. Encouraging participation in clubs, teams, and local activities expands social experience. Real-world interaction supports social skill development. Diverse relationships build adaptability.

The digital world will remain a significant part of children's lives. Preparing them to engage responsibly ensures long-term success. Guidance during formative years shapes habits and attitudes. Early structure influences future independence.

Ultimately, parenting beyond the screen is about cultivating presence. Attention, conversation, and shared experiences form the core of healthy development. Technology should complement family life, not dominate it. Intentional boundaries preserve balance.

By combining modeling, communication, structure, and community engagement, parents can guide children toward responsible digital use. The objective is resilience, awareness, and self-regulation. With thoughtful leadership, families can navigate the digital landscape while maintaining strong relational foundations.

CHAPTER . 33

The Art of Deep Work

Deep work is the capacity to concentrate without distraction on a cognitively demanding task. It is the discipline of sustained attention applied toward meaningful goals. In an age defined by interruption, deep work becomes both rare and valuable. It transforms information into insight and effort into mastery.

The modern environment is structured around rapid input and frequent switching. Notifications, messages, and alerts fragment attention throughout the day. While these tools promise efficiency, they often reduce uninterrupted thinking time. Deep work requires the opposite condition: protected focus.

When attention remains stable over extended periods, the brain engages in complex pattern recognition. Ideas connect across domains. Problems that seemed unclear become structured. Creativity emerges from sustained engagement rather than scattered interaction.

The practice of deep work begins with intentional scheduling. Designating specific blocks of time for focused activity establishes boundaries. During these periods, distractions are minimized. Consistency reinforces the habit of concentration.

Environmental design plays a significant role. A quiet workspace, limited visual clutter, and reduced digital interruptions support immersion. Physical conditions influence mental states. When surroundings signal focus, the mind adapts accordingly.

Deep work benefits from clear objectives. Defined goals guide attention and prevent drift. Breaking large projects into structured milestones enhances momentum. Clarity reduces cognitive overload.

Discipline is strengthened through repetition. Like physical training, attention improves with practice. Gradual increases in focus duration build endurance. Over time, the capacity for sustained engagement expands.

Technology can either hinder or support deep work. When used indiscriminately, devices introduce fragmentation. However, focus tools such as application blockers and offline modes can reduce interference. Intentional configuration turns devices into allies.

Deep work also enhances professional competence. Complex tasks require uninterrupted reasoning. Writing, analysis, engineering, and design all benefit from extended concentration. Quality improves when depth replaces haste.

In educational contexts, deep work supports mastery. Students who allocate time for uninterrupted study often retain information more effectively. Reflection solidifies learning. Concentration strengthens comprehension.

Psychologically, deep engagement fosters fulfillment. Completing challenging tasks generates a sense of accomplishment. Progress builds confidence. Achievement derived from sustained effort tends to be more satisfying than fleeting stimulation.

Deep work contrasts with reactive activity. Constantly responding to messages and notifications keeps attention externally directed. While responsiveness has value, balance is necessary. Deliberate creation requires inward focus.

Establishing rituals can help initiate deep work sessions. Consistent start times, designated spaces, and preparatory routines signal the brain to concentrate. Ritual reduces transition friction. Predictability enhances efficiency.

Time management strategies such as batching similar tasks also support depth. Grouping communication into specific windows prevents continuous interruption. This structure protects cognitive continuity. Efficiency increases as switching decreases.

Rest is an integral companion to deep work. Focused effort must be balanced with recovery. Periods of relaxation allow neural consolidation. Sustainable productivity depends on cycles of engagement and rest.

Social expectations may need adjustment to accommodate deep work. Organizations can encourage uninterrupted work periods. Cultural norms that respect focus time improve collective performance. Shared understanding reinforces discipline.

Deep work contributes to innovation. Breakthrough ideas often arise from extended contemplation. Complex systems require layered analysis. Sustained thought reveals relationships that brief attention cannot uncover.

The art of deep work also involves resisting the temptation of constant novelty. The digital environment offers endless stimulation. Choosing depth over distraction is an intentional act. It reflects prioritization of long-term goals.

Measurement of progress can reinforce commitment. Tracking completed focused sessions provides feedback. Incremental improvements build confidence. Awareness of progress sustains motivation.

Deep work enhances personal growth. Skills develop more rapidly under concentrated practice. Expertise emerges through deliberate repetition. Focused effort accelerates learning curves.

Community support can encourage depth. Work environments that minimize unnecessary interruptions promote collective focus. Shared commitment to concentration fosters productivity. Collaboration benefits from individual preparation.

In a world structured around engagement metrics, deep work represents a counterbalance. It shifts emphasis from visibility to substance. It values outcomes over reaction. It restores control over attention.

The neon systems described earlier thrive on distraction. Deep work interrupts that pattern. By reclaiming sustained concentration, individuals reduce dependence on external stimulation. Focus becomes self-directed.

Practicing deep work requires patience. Initial sessions may feel challenging. Discomfort diminishes with repetition. Gradual adaptation strengthens mental endurance.

The benefits extend beyond professional output. Deep engagement cultivates clarity of thought. It improves decision-making. It strengthens intellectual confidence.

Deep work ultimately reconnects individuals with agency. When attention is intentionally directed, productivity aligns with purpose. Effort becomes meaningful. Achievement reflects deliberate choice.

In becoming proficient at deep work, we reclaim one of the most powerful human capabilities: the ability to think deeply, create intentionally, and act with sustained focus. This discipline forms a cornerstone of becoming human again in a distracted world.

CHAPTER . 34

Rituals of Disconnection

In a world designed for constant connection, deliberate disconnection becomes a form of renewal. Technology encourages continuous engagement, yet the human mind and body require periodic rest from stimulation. Without structured pauses, cognitive and emotional fatigue accumulate. Rituals of disconnection restore balance.

A ritual differs from a random break. It is intentional, consistent, and symbolic. It signals transition from one state to another. When applied to technology use, rituals create psychological boundaries between engagement and rest. This structure reinforces control rather than compulsion.

One common ritual is the digital sabbath. Setting aside specific hours or days without nonessential device use allows the nervous system to recalibrate. During these periods, attention returns to physical surroundings and interpersonal interaction. The absence of notifications reduces background stress.

Rituals of disconnection can begin with simple practices. Turning off devices at a designated time each evening establishes predictable separation. Charging phones outside the bedroom reinforces sleep hygiene. Clear temporal boundaries protect rest.

Environmental cues strengthen these rituals. Placing devices in a specific location signals intentional disengagement. Replacing screen time with alternative activities supports habit formation. Reading, walking, journaling, or conversation can fill the space constructively.

The psychological benefits of structured disconnection are significant. Reduced exposure to continuous stimuli lowers stress levels. The brain experiences relief from constant monitoring demands. Calm environments support emotional regulation.

Disconnection also enhances presence in relationships. When devices are absent, conversations deepen. Eye contact increases. Listening improves. Attention becomes fully directed toward the person in front of us.

Family-based rituals are especially powerful. Shared technology-free meals create consistent opportunities for interaction. Regular device-free evenings encourage bonding. Predictable routines help children and adults alike adjust to boundaries.

Workplace rituals can also support sustainable productivity. Scheduled focus periods without messaging interruptions enhance concentration. Clear expectations regarding response times reduce anxiety. Organizational norms that respect offline intervals improve morale.

Outdoor rituals of disconnection offer additional restoration. Spending time in natural environments without devices provides sensory variety. Natural settings reduce mental fatigue. Exposure to sunlight and movement strengthens physical health.

Reflection is a core component of disconnection rituals. Journaling or quiet contemplation during technology-free periods clarifies thoughts. Without external input, internal awareness increases. Insight often emerges in silence.

These rituals do not require complete abstinence from technology. Instead, they emphasize rhythm. Alternating between engagement and withdrawal maintains balance. Cycles of use and rest mirror natural biological processes.

Consistency transforms rituals into habits. Repeated practice reduces resistance. Over time, disconnection becomes a normal part of daily structure. Discipline evolves into stability.

Communities can support collective rituals. Organized events that encourage device-free participation normalize presence. Group commitments to periodic disconnection foster accountability. Shared experiences reinforce cultural change.

Digital overload is cumulative. Without intentional breaks, exposure to notifications, news, and content builds continuously. Ritualized pauses interrupt this accumulation. Regular resets prevent chronic overstimulation.

The neon environment thrives on uninterrupted engagement. Infinite feeds and continuous alerts discourage natural stopping points. Rituals reintroduce those boundaries deliberately. They restore agency over time.

Rituals of disconnection also enhance creativity. When input decreases, the mind begins to synthesize internally. Daydreaming and reflection support innovation. Silence allows subconscious processing.

Emotional clarity improves during structured breaks. Continuous exposure to social comparison or conflict can intensify anxiety. Temporary withdrawal reduces emotional reactivity. Perspective strengthens when stimuli decrease.

Designers can encourage disconnection by incorporating gentle exit points. Features such as session reminders or daily usage summaries support awareness. Encouraging breaks within interfaces aligns technology with well-being.

The effectiveness of rituals depends on intention. Casual attempts may fade quickly. Committed routines, however, can transform habits. Consistency builds resilience against constant stimulation.

Rituals also create anticipation. Knowing that a technology-free period will occur later can reduce the urge for constant checking. Predictable boundaries lower cognitive tension. Structure replaces uncertainty.

In educational settings, scheduled disconnection improves focus. Device-free study sessions enhance comprehension. Periodic breaks prevent burnout. Balanced pacing supports long-term learning.

At a societal level, cultural acceptance of disconnection normalizes rest. When taking time offline is respected, individuals feel less pressure to remain available. Social norms influence behavior significantly.

Rituals of disconnection do not represent rejection of technology. They represent integration. Technology remains useful when engaged intentionally. Periodic withdrawal ensures that engagement remains voluntary.

The practice reinforces autonomy. Choosing when to connect and disconnect restores decision-making power. Awareness of usage patterns strengthens self-regulation. Boundaries cultivate independence.

Over time, rituals reshape relationships with devices. Instead of default engagement, interaction becomes deliberate. Instead of compulsion, connection becomes choice. This shift transforms digital experience.

The benefits extend beyond individual well-being. Communities that adopt shared disconnection practices often report improved dialogue and cooperation. Reduced digital distraction enhances interpersonal attention. Collective focus strengthens bonds.

In a constantly connected world, silence becomes valuable. Structured pauses create space for reflection, creativity, and rest. Rituals make these pauses reliable rather than accidental.

By embedding disconnection into daily and weekly routines, individuals and communities reclaim equilibrium. Technology regains its role as tool rather than master. Rhythm replaces excess.

Rituals of disconnection are simple but powerful. They remind us that control over attention is possible. They demonstrate that engagement can coexist with rest. And they reinforce the principle that balance sustains both productivity and well-being.

Chapter . 35

Mindfulness in the Machine Age

Mindfulness is the practice of paying deliberate attention to the present moment without judgment. It is not withdrawal from the world, but a deeper engagement with it. In the machine age, where stimuli arrive continuously and rapidly, mindfulness becomes an essential stabilizing skill. It restores awareness in environments designed for distraction.

Modern systems are built to capture attention automatically. Notifications, feeds, and recommendation engines operate in the background of daily life. These mechanisms encourage reactive behavior rather than reflective response. Mindfulness interrupts that automatic cycle.

At its core, mindfulness creates a pause between stimulus and action. In that pause, choice becomes possible. Without awareness, behavior is often driven by habit or impulse. With awareness, individuals can decide intentionally how to respond.

The machine age increases cognitive load. Constant information flow can overwhelm the nervous system. Multitasking and rapid switching reduce clarity. Mindfulness counteracts this fragmentation by anchoring attention in a single experience at a time.

Breathing awareness is one of the simplest mindfulness practices. By observing the breath, individuals cultivate focus without external tools. This internal anchor is always available. It does not depend on technology or environment.

Mindfulness also strengthens emotional regulation. When emotions arise, conscious observation allows space for processing. Instead of reacting immediately, individuals can acknowledge feelings and choose thoughtful responses. This reduces impulsive decisions.

Digital environments often amplify emotional intensity. Content designed for engagement may provoke strong reactions. Mindfulness helps users recognize emotional triggers. Awareness reduces susceptibility to manipulation.

Practicing mindfulness does not require extended time commitments. Short moments of intentional attention throughout the day can be effective. Even brief pauses before checking a device can restore balance. Consistency matters more than duration.

Machine-age environments reward speed. Rapid responses are valued in many professional and social contexts. While efficiency is important, constant acceleration can lead to stress. Mindfulness introduces deliberate pacing.

Attention training is a key benefit of mindfulness. Regular practice strengthens the ability to sustain focus. Over time, distractions lose some of their pull. Concentration becomes more stable.

Mindfulness also improves decision-making. When individuals observe thoughts without immediate identification, clarity increases. Emotional reactions become data rather than directives. This perspective enhances judgment.

In workplaces, mindfulness can reduce burnout. Structured pauses, reflective meetings, and focused work periods improve well-being. Organizations that support mindful practices often see increased productivity and morale. Balanced environments sustain performance.

Educational systems benefit from integrating mindfulness techniques. Students who practice awareness demonstrate improved concentration and reduced anxiety. Short guided sessions can prepare learners for deep engagement. Attention becomes more manageable.

Mindfulness complements rituals of disconnection. While disconnection reduces external stimulation, mindfulness refines internal awareness. Together, they strengthen attentional control. External quiet and internal clarity reinforce each other.

In a machine-driven world, many processes are automated. Algorithms handle tasks ranging from recommendations to logistics. Mindfulness ensures that humans remain conscious participants rather than passive observers. It preserves agency within automated systems.

Practicing mindfulness can involve simple techniques such as body scans, attentive listening, or focused observation of surroundings. These exercises ground awareness in immediate experience. Sensory attention reduces mental drift.

Stress reduction is one of the most widely documented outcomes of mindfulness practice. Lower stress levels contribute to better health and cognitive performance. Calm states improve resilience in demanding environments.

The machine age introduces constant performance measurement. Metrics track productivity, engagement, and output. While data can inform improvement, excessive focus on evaluation may create pressure. Mindfulness shifts attention from external measurement to internal awareness.

By cultivating present-moment focus, individuals reduce rumination about past events or future uncertainties. This stabilization supports mental health. Balanced awareness fosters adaptability.

Mindfulness also enhances creativity. When the mind is less cluttered by distraction, novel connections emerge. Observing thoughts without judgment allows ideas to develop naturally. Creativity thrives in spacious awareness.

Community-based mindfulness initiatives can amplify impact. Group meditation sessions, workshops, and shared practices normalize awareness. Collective participation reinforces individual commitment. Social support enhances consistency.

Technology can support mindfulness when used intentionally. Guided meditation applications and reminder tools can facilitate practice. However, these tools should serve awareness rather than replace it. The goal remains internal focus.

Mindfulness is not about eliminating thought. It is about observing thought without becoming controlled by it. In the machine age, where algorithms influence preferences and habits, this distinction is crucial. Awareness protects autonomy.

Developing mindfulness requires patience. Like any skill, it improves with repetition. Regular practice gradually changes attention patterns. Over time, awareness becomes more automatic.

The benefits extend beyond personal well-being. Mindful individuals contribute to calmer conversations and more thoughtful collaboration. Reduced impulsivity enhances social harmony. Awareness improves collective interaction.

In environments shaped by speed and automation, mindfulness reintroduces human-centered pacing. It emphasizes reflection before reaction. It restores equilibrium between input and interpretation.

By cultivating mindfulness in the machine age, individuals strengthen resilience against distraction, stress, and manipulation.

Awareness becomes a stabilizing force within complex systems. Presence replaces reactivity.

Mindfulness does not oppose technology. It ensures that technology serves conscious users. With practice, attention becomes steady, intentional, and grounded. In doing so, we reinforce our capacity to navigate digital environments with clarity and balance.

CHAPTER . 36

Building Local Resilience

Resilience is the capacity of a community to withstand disruption and recover from challenges. In a world shaped by global networks, local resilience remains essential. While digital systems connect people across vast distances, stability often depends on strong nearby relationships and resources. Communities that are locally grounded respond more effectively to uncertainty.

Local resilience begins with relationships. When neighbors know one another, cooperation becomes natural. Trust reduces friction during times of stress. Informal support networks often provide faster assistance than distant institutions.

Strong local economies contribute to stability. Supporting neighborhood businesses keeps resources circulating within the community. Economic activity anchored locally creates interdependence. This interdependence strengthens collective security.

Food systems are a central component of resilience. Community gardens, local farms, and regional supply chains reduce reliance on extended distribution networks. Shorter supply chains can improve reliability during disruptions. Access to locally produced goods enhances adaptability.

Energy resilience also matters. Communities that diversify energy sources and invest in sustainable infrastructure reduce vulnerability. Localized energy solutions can provide stability during broader system interruptions. Planning ahead increases preparedness.

Education contributes to local strength. Schools that partner with community organizations create shared responsibility for development. Lifelong learning initiatives strengthen workforce adaptability. Knowledge sharing within communities builds capacity.

Health infrastructure is another pillar. Accessible clinics, preventative care programs, and public health education enhance well-being. Strong local health systems respond quickly to emerging concerns. Community awareness supports collective safety.

Social cohesion plays a vital role in resilience. Regular gatherings, cultural events, and collaborative projects reinforce connection. Shared experiences cultivate belonging. Belonging strengthens cooperation.

Communication networks should complement local systems. While digital tools enable coordination, in-person communication remains foundational. Community meetings and public forums foster transparency. Clear dialogue reduces misunderstanding.

Disaster preparedness illustrates the importance of local resilience. Emergency plans, resource sharing agreements, and volunteer coordination improve response capability. Communities that prepare together recover more efficiently. Planning builds confidence.

Civic engagement strengthens resilience as well. Participation in local governance ensures that decisions reflect community priorities. Active involvement increases accountability. Engagement fosters ownership.

Infrastructure maintenance supports stability. Well-maintained roads, public spaces, and utilities reduce risk. Regular investment in communal assets prevents deterioration. Shared responsibility ensures longevity.

Cultural preservation contributes to resilience. Traditions, languages, and local art reinforce identity. Cultural continuity strengthens social bonds. Identity supports unity during change.

Technology can enhance local resilience when used thoughtfully. Digital platforms can coordinate community events, distribute information, and organize volunteers. However, reliance should remain balanced. Technology should support, not replace, local interaction.

Financial resilience is also important. Community-based savings groups, cooperative lending models, and local investment initiatives can provide support during economic shifts. Diversified financial strategies increase flexibility. Preparedness reduces vulnerability.

Environmental stewardship enhances long-term resilience. Sustainable land use, conservation efforts, and responsible resource management protect natural assets. Healthy ecosystems support community stability. Environmental care strengthens future security.

Intergenerational cooperation reinforces resilience. Older residents contribute experience and mentorship, while younger members offer energy and innovation. Shared learning builds continuity. Collaboration bridges age groups.

Trust is the foundation of resilient systems. Transparency in local institutions fosters confidence. Reliable communication strengthens cooperation. Trust reduces fragmentation.

Building local resilience requires intentional effort. It does not occur automatically. It develops through consistent engagement and shared responsibility. Commitment over time produces durable networks.

Local resilience does not reject global connection. Instead, it complements it. Strong communities can participate confidently in global systems while maintaining stability at home. Balance between local grounding and global engagement ensures flexibility.

Education about resilience can empower residents. Workshops on emergency preparedness, financial planning, and community organization enhance readiness. Knowledge strengthens adaptability. Prepared individuals support collective strength.

Shared problem-solving reinforces unity. When communities address challenges together, bonds deepen. Collaborative action transforms obstacles into opportunities for cooperation. Collective effort builds confidence.

Local resilience also supports mental health. Strong social networks reduce isolation. Regular interaction fosters belonging and support. Emotional security contributes to overall well-being.

Public spaces are critical to resilience. Parks, libraries, community centers, and meeting halls provide gathering points. Accessible spaces encourage participation. Shared environments facilitate interaction.

Volunteer culture strengthens adaptability. When residents contribute time and skills, communities become more self-sufficient. Service reinforces responsibility. Engagement enhances solidarity.

Building local resilience requires ongoing evaluation. Communities can assess strengths and identify gaps. Continuous improvement ensures relevance. Reflection supports growth.

The neon systems of global connectivity are powerful, but local structures provide grounding. When individuals feel connected to their immediate surroundings, they gain stability. Proximity fosters trust and cooperation.

By investing in relationships, infrastructure, education, and shared resources, communities create durable foundations. Local resilience ensures that technological change does not erode stability. It transforms vulnerability into preparedness.

In a rapidly changing digital world, resilience begins at home—within neighborhoods, towns, and cities. Strong local networks provide support during uncertainty and enable communities to thrive alongside global systems. Resilience is not isolation; it is rooted connection.

Chapter . 37

Economic Models for Human Flourishing

Economic systems shape more than markets. They influence behavior, opportunity, well-being, and social stability. The structure of incentives determines what is rewarded and what is overlooked. If we want societies to flourish, we must examine not only growth, but the quality of that growth.

Traditional economic models often prioritize efficiency, productivity, and expansion. These metrics can generate innovation and material progress. However, when measured narrowly, they may overlook human factors such as mental health, community strength, environmental sustainability, and long-term resilience. Flourishing requires a broader definition of success.

An economic model for human flourishing begins with values. It asks what outcomes matter most for individuals and communities. It considers dignity, opportunity, fairness, and stability as essential components. It aligns financial incentives with social well-being.

One approach emphasizes stakeholder participation. Rather than focusing solely on shareholder returns, organizations can account for employees, customers, communities, and ecosystems. This perspective encourages balanced decision-making. It distributes responsibility across all affected groups.

Another model highlights cooperative ownership. Worker cooperatives and community-owned enterprises allow participants to share in both decision-making and profits. Shared ownership can increase engagement and accountability. It strengthens local economic ties.

Economic flourishing also depends on fair labor practices. Stable wages, safe working conditions, and access to benefits contribute to security. When workers feel valued and protected, productivity often improves. Economic dignity supports social stability.

Innovation remains important within flourishing models. However, innovation should serve human needs rather than simply maximize engagement or consumption. Technologies can be designed to enhance quality of life. Purpose-driven development encourages sustainability.

Digital economies, in particular, benefit from reevaluating metrics. If platforms optimize solely for attention, outcomes may include distraction and polarization. Introducing well-being indicators alongside engagement metrics can rebalance priorities. Measurement shapes behavior.

Income distribution plays a role in societal health. Excessive inequality can limit opportunity and reduce social mobility. Policies that support equitable access to education, healthcare, and capital can broaden participation. Inclusive growth strengthens collective prosperity.

Economic models for flourishing also consider environmental impact. Sustainable resource use ensures long-term viability. Integrating ecological accounting into financial systems encourages responsible production. Economic success should not compromise future generations.

Community-based finance initiatives offer alternative pathways. Local investment funds, microfinance programs, and cooperative credit systems empower residents. These models keep capital circulating within communities. Financial inclusion fosters resilience.

Education and workforce development are central components. Training programs that adapt to evolving technological landscapes enable continuous participation. Lifelong learning aligns human skills with economic change. Adaptability ensures relevance.

Well-being metrics are increasingly recognized in policy discussions. Measures such as life satisfaction, health outcomes, and social trust provide complementary perspectives to traditional GDP calculations. Broader indicators capture aspects of human experience beyond output alone. Data can guide humane policy.

Entrepreneurship within flourishing models can emphasize social impact. Mission-driven businesses address community needs while maintaining financial sustainability. Social enterprises demonstrate that profit and purpose can coexist. Innovation can be aligned with ethics.

Economic resilience benefits from diversification. Communities with varied industries and revenue streams are better positioned to handle disruptions. Dependence on a single sector increases vulnerability. Diversity strengthens stability.

Digital platforms can support flourishing when designed ethically. Marketplaces that empower small businesses and creators can expand opportunity. Transparent algorithms and fair fee structures enhance trust. Technology becomes an enabler rather than a gatekeeper.

Public policy plays a crucial role in shaping economic direction. Tax structures, regulatory frameworks, and social programs influence distribution and incentives. Thoughtful governance ensures alignment with shared values. Institutions guide outcomes.

Collaboration between public and private sectors can generate balanced solutions. Partnerships that address infrastructure, education, and sustainability promote long-term benefits. Shared responsibility reduces fragmentation. Cooperation enhances effectiveness.

Flourishing economies prioritize human capital. Investment in health, education, and skills development yields substantial returns. When individuals thrive, productivity and innovation follow naturally. Well-being and growth reinforce one another.

Time is also an economic resource. Work-life balance contributes to personal fulfillment. Flexible arrangements can improve satisfaction and efficiency. Recognizing the value of time supports holistic prosperity.

Cultural vitality is another component of flourishing. Supporting arts, creativity, and local traditions enriches community life. Cultural industries contribute both economically and socially. Expression strengthens identity.

Transparency within financial systems enhances trust. Clear reporting, ethical governance, and accountability mechanisms reduce corruption risks. Trust in institutions supports stable markets. Integrity sustains progress.

Economic models for human flourishing acknowledge interdependence. Individuals, communities, businesses, and ecosystems are connected. Decisions in one domain influence others. Systems thinking improves policy design.

The goal is not to eliminate competition, but to frame it within ethical boundaries. Healthy competition can drive innovation and efficiency. However, competition should not undermine fairness or well-being. Balanced incentives produce sustainable outcomes.

Flourishing requires long-term perspective. Short-term gains must be evaluated in relation to lasting impact. Investment strategies that consider decades rather than quarters encourage stability. Patience fosters resilience.

In the digital age, data itself becomes a form of economic capital. Ensuring that data practices respect privacy and agency is essential. Responsible data governance aligns technological progress with human rights. Ethical standards support flourishing.

Ultimately, economic models for human flourishing integrate material prosperity with social, psychological, and environmental health. They redefine success to include dignity, equity, sustainability, and opportunity. Growth remains valuable, but it is guided by purpose.

By redesigning incentives and broadening metrics, societies can create systems that reward contribution, cooperation, and innovation while protecting well-being. Economic structures can evolve to serve humanity more fully. Flourishing becomes the measure of progress.

CHAPTER . 38

Technology as Tool, Not Tyrant

Technology is one of humanity's most powerful achievements. From the earliest tools to modern digital systems, innovation has extended our capabilities and expanded our possibilities. Tools amplify human intention. They allow us to accomplish tasks that would otherwise be impossible or inefficient.

The distinction between a tool and a tyrant lies in control. A tool responds to human direction. A tyrant dictates behavior. When technology serves human goals, it remains within its intended role. When systems begin to shape behavior without conscious oversight, balance can shift.

Modern digital environments are designed with sophisticated feedback loops. Algorithms optimize engagement, personalize experiences, and predict preferences. These systems can be beneficial when aligned with user intent. However, without intentional boundaries, they may gradually influence habits and decisions.

Maintaining technology as a tool requires clarity of purpose. Before engaging with a platform or device, it helps to ask what objective it serves. If usage supports productivity, connection, learning, or creativity, it fulfills a constructive role. Awareness ensures alignment.

Design choices significantly influence whether technology functions as a servant or a driver. Transparent settings, user-controlled customization, and clear data policies empower individuals. When users retain meaningful control, agency is preserved. Empowerment keeps systems accountable.

Personal boundaries are equally important. Time limits, focused usage periods, and intentional disconnection reinforce control. Structured engagement prevents drift into habitual overuse. Discipline transforms interaction into deliberate action.

Organizations can reinforce this principle by designing systems that prioritize user well-being. Ethical defaults, minimal data collection, and clear opt-in features support autonomy. When systems respect attention and privacy, trust increases. Responsible design sustains long-term adoption.

Education strengthens the ability to treat technology as a tool. Understanding how algorithms function enables informed choices. Digital literacy reduces dependence on automatic recommendations. Knowledge restores perspective.

In workplaces, technology should enhance human skill rather than replace it indiscriminately. Automation can remove repetitive burdens while preserving meaningful roles. Collaboration between humans and machines maximizes strengths on both sides. Partnership produces better outcomes than substitution alone.

Technology as a tool also implies reversibility. Users should be able to step away without penalty. Systems that trap engagement undermine autonomy. Exit options preserve freedom.

Designing for portability supports this principle. When users can move data between platforms easily, dependence decreases. Interoperability encourages competition and choice. Flexibility maintains balance.

Cultural attitudes influence whether technology dominates or supports life. When society values constant connectivity above rest and presence, systems adapt accordingly. When balance is prioritized, design trends shift. Collective norms shape development.

Mindful consumption reinforces the tool mindset. Using devices intentionally, rather than reflexively, maintains clarity. Conscious interaction transforms digital environments into resources. Intention guides behavior.

Leadership within technology companies plays a significant role. Executives and engineers who prioritize ethical standards can influence product direction. Organizational values shape outcomes. Responsibility at the top affects design at every level.

Public dialogue about technology's role is essential. Open discussions about benefits and risks encourage transparency. Informed debate prevents extremes of either blind acceptance or total rejection. Balanced conversation supports thoughtful progress.

Regulatory frameworks can help maintain technology's role as a tool. Clear guidelines on privacy, competition, and consumer protection establish boundaries. Oversight ensures systems remain accountable. Governance supports stability.

However, individual responsibility remains central. Even within regulated environments, users make daily choices. Habits determine experience. Consistent awareness preserves autonomy.

The metaphor of tool versus tyrant extends beyond devices. It reflects a broader relationship between humans and systems. Any structure—economic, political, or technological—can either support or constrain. Intent and design determine direction.

To keep technology in its proper role, periodic evaluation is necessary. Reflecting on usage patterns reveals whether systems serve goals or distract from them. Adjustments restore alignment. Reflection sustains control.

Balance does not require minimalism, but intentionality. Technology can enrich creativity, collaboration, and learning. It can connect distant communities and expand access to information. Its value depends on application.

The neon systems described throughout this book are powerful infrastructures. Yet infrastructure should remain subordinate to human values. When technology supports dignity, autonomy, and flourishing, it fulfills its purpose. When it overrides these principles, recalibration is needed.

Maintaining technology as a tool requires ongoing commitment from individuals, institutions, and designers. Shared responsibility ensures alignment. Continuous awareness preserves equilibrium.

By consistently choosing intentional engagement, societies can harness innovation without surrendering agency. Technology can remain an instrument of progress rather than a force of domination. The distinction depends on vigilance, design, and culture.

When tools are guided by conscious users and ethical frameworks, they expand possibility without diminishing freedom. This balance ensures that technological advancement strengthens humanity rather than redefining it.

CHAPTER . 39

Restoring Shared Reality

Shared reality is the foundation of cooperation. For societies to function, people must agree on basic facts, common references, and reliable sources of information. While disagreement is natural and healthy in many contexts, a shared baseline of truth enables constructive dialogue. Without it, coordination becomes fragile.

In the digital age, informational environments are increasingly personalized. Algorithms curate content based on behavior, preferences, and engagement history. As a result, individuals may encounter different narratives, priorities, and interpretations of events. This personalization can narrow exposure to diverse viewpoints.

When people inhabit separate informational streams, consensus becomes more difficult. Even when discussing the same issue, participants may reference different data or framing. This divergence can create misunderstanding. Dialogue weakens when foundational assumptions differ.

Restoring shared reality begins with reinforcing reliable information channels. Trustworthy journalism, transparent institutions, and evidence-based research provide anchors for public understanding. Clear sourcing and accountability strengthen credibility. Verification processes help maintain standards.

Media literacy is central to this restoration. Individuals who understand how content is produced, distributed, and amplified are better equipped to evaluate claims. Recognizing persuasive techniques reduces vulnerability. Critical evaluation supports informed judgment.

Transparency from platforms also contributes to shared reality. Clear labeling of sponsored content, political advertising, and algorithmic recommendations reduces ambiguity. When users understand why they are seeing certain information, trust improves. Visibility supports accountability.

Fact-checking institutions play a supportive role. Independent review of claims can correct misinformation. Rapid response mechanisms help limit the spread of false narratives. However, corrections must be visible and accessible to be effective.

Restoring shared reality also requires fostering respectful dialogue. Civil discussion allows disagreement without hostility. Structured forums that encourage evidence-based conversation can bridge divides. Listening as well as speaking strengthens understanding.

Educational systems are key contributors. Teaching students how to assess sources, compare perspectives, and analyze arguments prepares them for complex environments. Skills in logic and reasoning reduce reliance on emotional cues alone. Education strengthens epistemic resilience.

Technology can both disrupt and support shared reality. While algorithmic amplification can spread sensational content, design choices can prioritize credible sources. Platforms can highlight context, encourage diverse exposure, and reduce misinformation visibility. Responsible configuration influences outcomes.

Community-level engagement enhances shared understanding. Local discussions, public meetings, and collaborative initiatives encourage face-to-face exchange. In-person dialogue reduces misinterpretation. Shared experiences build trust.

Restoring shared reality does not require uniformity of opinion. Diversity of thought is valuable and necessary. The goal is not agreement on every issue, but alignment on verifiable facts. Common ground enables productive debate.

Fact-based discourse benefits from transparency in data collection and research methods. When studies are open to review and replication, confidence increases. Openness supports integrity. Integrity sustains trust.

Cultural narratives also influence shared perception. Storytelling shapes understanding of events and values. Encouraging accurate representation in media strengthens alignment between story and evidence. Balanced narratives reduce distortion.

Digital systems should support context rather than fragment it. Providing background information alongside headlines improves comprehension. Linking to primary sources enhances depth. Contextualization strengthens clarity.

Individual responsibility contributes as well. Before sharing information, verifying its accuracy helps prevent misinformation spread. Deliberate communication reduces noise. Conscious sharing supports collective stability.

Shared reality benefits from inclusivity. When diverse voices are represented in public discourse, perspectives broaden. Inclusion fosters legitimacy. Representation strengthens trust in institutions.

Algorithmic transparency can be complemented by user choice. Allowing individuals to adjust recommendation settings promotes exposure diversity. Customization supports balanced consumption. Control enhances awareness.

Restoration efforts should avoid censorship that undermines legitimate expression. Protecting free speech while addressing harmful misinformation requires careful design. Clear standards and consistent enforcement maintain fairness. Balance is essential.

Community media initiatives can reinforce local trust. Independent journalism at regional levels provides relevant information. Local reporting strengthens connection to immediate environments. Proximity increases accountability.

Digital literacy campaigns help citizens understand manipulation techniques. Awareness of deepfakes, coordinated campaigns, and engagement-based amplification reduces confusion. Education empowers discernment. Knowledge stabilizes perception.

Restoring shared reality also involves cultivating empathy. Understanding that others may receive different information environments encourages patience. Recognizing structural influences reduces personal blame. Dialogue improves when participants acknowledge context.

Institutional trust grows when transparency is consistent. Open communication about decisions, policies, and data practices builds confidence. Accountability mechanisms ensure reliability. Trust develops through demonstrated integrity.

In the broader context of this book, restoring shared reality addresses challenges created by fragmented information ecosystems. Without common references, democratic participation weakens. With shared understanding, collaboration strengthens.

The process requires cooperation among individuals, communities, technology platforms, and policymakers. Each plays a role in maintaining clarity. Collective effort preserves stability.

Restoring shared reality is ultimately about commitment to truth-seeking. It involves valuing evidence over speculation and verification over assumption. It encourages dialogue grounded in facts. It supports unity without suppressing diversity.

When societies invest in transparency, education, and responsible design, they rebuild common understanding. Shared reality does not eliminate disagreement, but it provides a foundation for it. That foundation enables progress, cooperation, and trust in an increasingly complex world.

CHAPTER . 40

The New Covenant with the Future

Every civilization eventually faces a moment of reflection. The tools it creates begin to shape its identity, its priorities, and its destiny. At such moments, societies must decide whether their systems will define them—or whether they will redefine their systems. This is the essence of a covenant: a deliberate agreement about how we will move forward together.

The digital age has given humanity extraordinary power. We can communicate instantly across continents, automate complex tasks, simulate reality, and access vast knowledge within seconds. These capabilities expand opportunity. They also introduce responsibility. Power without intention can drift. Power with clarity can serve.

The "afterworld" described throughout this book is not an end state. It is a transitional environment shaped by incentives, algorithms, economic models, and attention structures. It reflects design decisions made over time. Because it is constructed, it can be reconstructed. Because it is human-made, it can be redesigned.

The new covenant with the future begins with awareness. We must understand how attention is influenced, how data is collected, how algorithms prioritize content, and how economic incentives shape outcomes. Awareness transforms passive participation into conscious engagement. Knowledge restores agency.

Next comes intentional design. Technology should be evaluated not only for capability, but for consequences. Systems can be built to support well-being, transparency, privacy, fairness, and sustainability. Design choices can align innovation with human flourishing. Ethical architecture is possible.

The covenant also requires shared responsibility. Individuals must cultivate attention, literacy, and mindful engagement. Families must guide children beyond screens. Communities must rebuild local resilience. Institutions must regulate transparently and fairly. No single group can carry the burden alone.

Education is central to this agreement. Schools must teach critical thinking, digital literacy, deep focus, and ethical reasoning. Students must learn how systems operate so they can participate intelligently in shaping them. Depth of understanding is the foundation of long-term stability.

Economic systems must evolve to reward contribution, not merely extraction. Incentives should support value creation, cooperation, and sustainability. Metrics must expand beyond engagement and growth to include well-being and equity. Prosperity should be measured in human terms as well as financial ones.

Technology companies play a vital role in this covenant. Platforms can prioritize transparency, user control, privacy protection, and ethical defaults. Algorithms can be designed to promote diverse exposure rather than narrow reinforcement. Engagement can coexist with responsibility.

Governments must ensure fair competition, protect civil rights, and maintain democratic integrity. Regulation should guide innovation toward public benefit. Oversight should be adaptive, informed, and balanced. Governance provides structure for accountability.

Citizens must remain active participants. Democracy in the digital era depends on engagement that is informed and respectful. Shared reality must be preserved through commitment to evidence and dialogue. Civic participation strengthens collective direction.

The covenant also involves culture. Societies can choose to value presence over constant stimulation, depth over distraction, and cooperation over division. Cultural norms influence daily behavior. When communities prioritize balance, systems adjust accordingly.

Environmental responsibility is another pillar. The future depends on sustainable resource use and ecological stewardship. Technological advancement should not compromise planetary stability. Long-term flourishing requires respect for natural systems.

At the individual level, the covenant encourages disciplined attention, mindful technology use, deep work, rituals of disconnection, and offline community engagement. These practices restore balance between digital and physical life. Human agency becomes central rather than reactive.

The future will likely include increasingly powerful technologies—advanced artificial intelligence, immersive virtual environments, and global data networks. These tools can amplify creativity, health, education, and collaboration. Their impact depends on alignment with shared values.

The covenant does not demand perfection. It calls for continuous improvement. Systems will evolve. Challenges will emerge. But with awareness and cooperation, adaptation can remain intentional.

This agreement with the future is fundamentally about stewardship. We are temporary custodians of powerful systems. The decisions made today shape experiences for generations ahead. Responsibility extends beyond immediate convenience.

By naming the systems that influence us, reclaiming attention, rebuilding community, redesigning education, regulating wisely, and prioritizing ethical design, we transform the digital landscape. The neon glow becomes illumination rather than distortion. Infrastructure becomes supportive rather than controlling.

The covenant envisions technology as servant, economy as inclusive, education as deep, community as resilient, and attention as sacred. It integrates innovation with humanity. It harmonizes progress with wisdom.

The future is not predetermined by algorithms or markets. It is shaped by choices. Each boundary set, each design improved, each conversation grounded in truth contributes to direction. Collective action determines outcome.

This book began by examining the rise of digital systems that influence desire, identity, and power. It progressed through their psychological, political, and cultural effects. It concluded with pathways for restoration and balance. The arc moves from recognition to responsibility.

The new covenant with the future is a commitment to alignment—between technology and ethics, economy and equity, speed and depth, connection and presence. It is an agreement to build systems that elevate human potential rather than diminish it.

If we honor this covenant, the afterworld becomes not a dystopia, but a transition toward a humane digital civilization. The neon gods lose their illusion of autonomy. They become tools—designed, guided, and governed by conscious societies.

The future will belong to those who understand systems and choose to shape them wisely. By embracing awareness, integrity, and collaboration, we ensure that innovation remains a force for flourishing. This is the promise of the new covenant: to transform power into responsibility, and technology into a partner in human progress.

Epilogue

The neon glow does not have to signal distortion. It can signal clarity.

Technology is neither hero nor villain. It is a mirror of human intention, encoded in design and sustained by behavior. The systems we inhabit reflect the priorities we reward. If engagement is rewarded, engagement grows. If well-being is rewarded, well-being grows. Systems follow incentives.

Throughout this journey, we explored attention, identity, power, education, economy, culture, and community. Each chapter revealed a common thread: structures shape experience. When we understand the structure, we gain leverage.

The path forward is not about retreating from progress. It is about aligning progress with wisdom. It is about ensuring that innovation enhances depth rather than replacing it. It is about building tools that serve humanity—clearly, ethically, and transparently.

The future is still unwritten.

Shared reality can be restored. Attention can be reclaimed. Communities can be rebuilt. Education can be redesigned. Technology can be guided. Economic systems can evolve. Regulation can adapt. Culture can remain diverse.

None of this requires perfection. It requires participation.

The neon gods only hold power if we forget that we built them.

Awareness returns authorship to us.

The afterworld becomes not an ending—but a beginning.

About the Author

Laura Lee is a writer, creative entrepreneur, and digital visionary whose work explores the intersection of technology, identity, and human meaning. Raised in a quiet town where storytelling was both tradition and lifeline, she developed an early appreciation for the power of language to shape connection, clarity, and transformation.

As the founder of a successful digital design agency, Laura has spent years helping individuals and organizations build thoughtful brands rooted in purpose and authenticity. Her professional work centers on intentional design—creating experiences that reflect not only visual excellence, but integrity and coherence.

Her writing carries the same philosophy. Blending reflection with practical insight, Laura explores how people can live consciously within rapidly evolving digital environments. She believes technology should enhance human creativity and dignity, not overshadow it.

For Laura, design is an act of responsibility. It is not only about aesthetics—it is about alignment. It is about ensuring that systems, stories, and structures reflect care for the people who interact with them.

Through her books, platforms, and creative consulting, she encourages others to reclaim their voice, strengthen their awareness, and build lives grounded in both innovation and authenticity.

This book reflects her ongoing commitment to conscious creation—inviting readers to think deeply, design wisely, and move forward with clarity in a world of constant change.

www.ingramcontent.com/pod-product-compliance
Lightning Source LLC
LaVergne TN
LVHW052318200926
848481LV00002B/596